W9-CSW-437

IMAGES OF ASIA

Fruits of South-East Asia

Titles in the series

At the Chinese Table
T. C. LAI

Balinese Paintings
A. A. M. DJELANTIK

Building a Malay House
PHILLIP GIBBS

Chinese Jade
JOAN HARTMAN-GOLDSMITH

Early Maps of South-East Asia
R. T. FELL

Folk Pottery in South-East Asia
DAWN F. ROONEY

Fruits of South-East Asia: Facts and Folklore
JACQUELINE M. PIPER

A Garden of Eden: Plant Life in South-East Asia
WENDY VEEVERS-CARTER

The House in South-East Asia
JACQUES DUMARÇAY

Indonesian Batik: Processes, Patterns and Places
SYLVIA FRASER-LU

The Kris: Mystic Weapon of the Malay World (2nd ed.)
EDWARD FREY

Macau
CESAR GUILLEN-NUÑEZ

Musical Instruments of South-East Asia
ERIC TAYLOR

Old Bangkok
MICHAEL SMITHIES

Riches of the Wild: Land Mammals of South-East Asia
EARL OF CRANBROOK

Sailing Craft of Indonesia
ADRIAN HORRIDGE

Sarawak Crafts
HEIDI MUNAN

Silverware of South-East Asia
SYLVIA FRASER-LU

Traditional Chinese Clothing
VALERY M. GARRETT

Fruits of South-East Asia

Facts and Folklore

JACQUELINE M. PIPER

SINGAPORE
OXFORD UNIVERSITY PRESS
OXFORD NEW YORK
1989

Oxford University Press

Oxford New York Toronto
Delhi Bombay Calcutta Madras Karachi
Petaling Jaya Singapore Hong Kong Tokyo
Nairobi Dar es Salaam Cape Town
Melbourne Auckland
and associated companies in
Berlin Ibadan

Oxford is a trade mark of Oxford University Press

ISBN 0 19 588904 5

Printed in Singapore by Kim Hup Lee Printing Co. Pte. Ltd.
Published by Oxford University Press Pte. Ltd.,
Unit 221, Ubi Avenue 4, Singapore 1440

For Ben and Sombat

Preface

THE original aim of this book was to describe something of the botany, horticulture, and uses of South-East Asian fruits, selecting a mere 30 or so from the far greater range of fruits grown in private gardens and orchards. However, as the book took shape, two other areas of interest clamoured increasingly for attention: on the one hand, the mythology and folklore associated with these fruits, their place in regional cultures in festivals, ceremonies and rituals, and handicraft design; on the other hand, the fascinating and enlightening accounts written by the first Western travellers to South-East Asia.

After these travellers, mainly statesmen and traders, came the naturalists, classifying the natural world, and after them, the anthropologists. The ethnobotanist is a latecomer in the field, investigating the links that exist between the plant life of an area and human social behaviour and customs: local use of native plants and local beliefs in the properties of plants. This book skirts the margins of ethnobotany, looking at fruits only and covering a wide region rather briefly. Nevertheless, it may indicate to the reader the wealth of this subject area in South-East Asia.

The introduction describes the broad environments of South-East Asia and the significance of these for fruit-growing. The ways in which fruits have spread from and into South-East Asia are recounted–the majority of fruits in this book are native to this corner of Asia but a few have been introduced from elsewhere. The main body of the book is made up of descriptions of the fruits with some details of their horticulture and growing characteristics (propagation, years to fruition, length of life, yields of fruit, etc.). The fruits described in this book—together with their parent plants—provide not only food but an enormous array of services and products, from the medical and the very practical to the cosmetic and superstitious. Some of the customs surrounding the production and uses of the fruits and any related folk-tales and beliefs are also described in these chapters, one of which deals with spices. Chapters 10 and 11 concentrate on the use of various fruits in art and design and in ceremony and rituals. The final chapter discusses briefly how fruits are propagated and grown commercially and details how some fruit trees may be grown at home from seed.

The text is illustrated with recent photographs, water-colours, and sketches; early engravings and water-colours of both the fruits and the South-East Asian scene complement these.

For making it possible to collect together this set of illustrations, my sincere thanks are due to Edith Jorge de De Martini (Didi) for donating her water-colours, to Sylvia Fraser-Lu for her photographs, to Garuda Indonesia for the photograph by Brian Morris, and to the Royal Botanic Gardens, Kew, the Board of Trustees of the Victoria and Albert Museum, London, the Tropenmuseum, Amsterdam, and the Vinmanmek Museum, Bangkok, for permission to use illustrations of material from their collections.

Also, my thanks to Carol Stratton and Miriam McNair Scott and to Urs Ramseyer for permission to reproduce illustrations from their books and to Wendy Veevers-Carter for the use of the drawings by Mohamed Anwar of Bogor. Finally, Ben Piper has provided not only several photographs but also a great deal of practical help with this book.

Oxford
April 1988

JACQUELINE M. PIPER

Contents

1
Introduction

EVER since man became a farmer rather than simply a gatherer of wild plants, he has influenced the crops he grows, improving their yields and changing their characteristics. Many plants have been transported from their place of origin to other regions and new growers have continued to select and breed the plants in their new locations. Movements of crops have occurred over very long periods of time: some crops were taken from India to China 4,000 years ago and others from Africa to India 3,000 years ago.

Fruits have been moved with the great tides of history: the rise of Islam took Seville oranges to Europe, the slave trade introduced breadfruit to the West Indies and mangoes to Africa, whilst the exploration and colonization of the New World and the East also brought new opportunities–Colombus's 'discovery' of America, and the subsequent linking of the Americas with world trade routes, enabled some plants to spread right across the tropics in suitable environments. From 1500 to 1665 a Portuguese sea route stretched from Portugal to the East via Brazil and the Cape of Good Hope; a Spanish galleon route ('the Manila ship') operated from the Philippines to Mexico between 1565 and 1815. The spice trade stimulated trading interest in the East, which attracted the colonial powers of the Netherlands, France, and Britain to the Orient. Once these foreign powers were established there, other products came to be traded, notably timbers such as teak, as well as gold and silk.

Next, the oriental fruits gradually began to be heard of in the West–curiosities such as the durian were frequently described in early travellers' tales, but it would be centuries before such fruits could be sampled away from the tropics. Where they could not bring back the fruits, travellers returned with drawings–some early lithographs are reprinted here from accounts by John Nieuhoff, who set sail from Amsterdam to the 'East Indies' in 1653; by de la Loubère, the 'Envoy extraordinary from the French King to the King of Siam'; by John Barrow, who visited Java in 1792 and 1793; and by others. In many cases, drawings dating from this early period were based on the traveller's sketches but elaborated upon by European engravers who had never visited the East.

The fruits in this book are in no way 'wild' or 'forest' fruits. They have been selected and bred by horticulturists in the region over hundreds of years. Like apples, strawberries, and peaches grown in the West, they show the fruit-grower's skill. Where the wild prototypes of these South-East Asian fruits are found, they are far removed in taste and size from the fruits on sale in markets.

These fruits are grown in orchards, plantations, and small farm plots. In every country of South-East Asia, most of the farming land is given over to small family farms, and more people depend on farming than on any other occupation. Pineapple, banana, coconut, and mango are widely grown in great plantations for canning or processing and export. The produce of small farmers is typically sold under a spreading sunshade at the roadside or in covered markets, as they have been for centuries: compare this book's cover, drawn in the nineteenth century, with Colour Plate 1, a present-day fruit stall.

All the fruits described here are tropical in origin, though some are now also grown in subtropical regions; the frost of cooler temperate regions would kill them. The trees respond differently to rainfall conditions: some require less water than others, and can survive long dry periods. In the lands under the monsoon, the rains are concentrated into one or two seasons of the year; during the intervening dry seasons, only a few light showers may occur over several months. Thus, for example, much of Thailand has a wet season from May to November and a dry season from December to May. Only crops that can survive 4 or 5 months without rain can grow there, unless they are irrigated. Some fruits require constantly available moisture: closer to the equator, Malaysia and Indonesia experience two monsoons a year–one from the south and one from the north–and dry periods are consequently short.

Temperature plays an important role in determining where crops are grown. Those fruits requiring slightly cooler conditions are grown in the hilly or mountainous areas, while some others prefer the steady temperatures of coastal locations. High winds, especially the typhoons which blow in from the Pacific, can restrict the areas in which fruit trees are grown.

Finally, soil type is also important. Sandy soils are usually drier and may not be suitable for trees which cannot withstand drought. Other trees hate waterlogging and cannot be grown on heavy, slow-draining soils. Some of the trees described in this book–like jackfruit, papaya, and breadfruit–produce very heavy yields of fruit each year and need very fertile soils. Other crops are equally happy on poorer soils and cashew, for example, actually produces a better crop on poor soils (on good soils, the tree produces abundant foliage, but fewer fruits).

2
Common Year-round Fruits

SEASONS are less well marked by temperature change in the tropics than in regions further from the equator and there are several plants which produce fruit all year round. Amongst these, four stand out as being most frequently offered in homes and hotels: banana, papaya, pineapple, and water-melon. The banana tree and the papaya tree are ubiquitous, occurring in twos and threes in gardens, as a few dozen in small farmer's plots, and in great multitudes in plantations. Water-melon and pineapple are both field crops, not tree crops; water-melon is grown on small plots as well as on large farms. Pineapple is most commonly grown in very large plantations for processing or canning for export, but great quantities are also eaten in South-East Asia.

Of these four common fruits, only banana is native to the region and has been cultivated here for millennia; water-melon probably reached South-East Asia about 1,000 years ago whilst pineapple and papaya are relative newcomers, introduced in the sixteenth century. A traditional crop, the banana (like rice) has helped to make Asian farming what it is. Asian societies are still very dependent on agriculture and bananas continue to contribute in many ways to local cultures and have a place in many customs. Many myths attach to the plant.

BANANA (*Musa* spp.)

At first glance, the banana plant is an odd sort of tree, growing to a height of 4–8 m, with an enormous hanging purple flower and great leaves accounting for almost half its total height; the huge bunches of fruit also seem quite out of proportion with its size (Plate 1). In fact, as it has no woody tissue, the banana plant is not really a tree at all, but a giant herb. Considered in those terms, like a giant lily, the size of the flowers and fruit seem more in keeping with the size of the plant.

Although bananas are now grown throughout the tropics, they are native to South-East Asia; the first records of the plant occur in the Epics of the Pali Buddhist canon (500–600 BC). Bananas were taken to Africa (probably from Indonesia) about

1. Six fruits, lithograph (de la Loubère, 1693).

1,500 years ago and from there to Arabia–the plant is referred to in the Koran as the Tree of Paradise–the Arabians knew something of the Asian tropics! In his classification of plants, Linnaeus gave bananas the names *Musa paradisica* (sweet bananas) and *Musa sapientum* (cooking bananas or plantains). *Musa* derives from the Arabic name, *mouz*, whilst *sapientum* may refer either to a mention by Pliny of Indian sages eating bananas or to the Christian tradition in which the banana is the tree of knowledge. In South-East Asia, banana trees are believed to be holy and are grown in temple grounds. Local Thai tradition has it that tall bananas are the home of a good guardian spirit: it is said that if, when the moon is full, one approaches the trees and prays, a beautiful lady with waist-length hair and a flower above her ear will appear. Farmers will not uproot this variety of banana from their land, in the belief that the spirit watches over them.

The banana plant is a perennial, growing from a rhizome which sends up shoots to form successive stems as each stem produces fruits then dies back. To attain its size, the banana must rely on very strong fibres within its leaves and stem; these fibres are used for tying and packaging. The banana 'trunk' is, in fact, the bases of the leaves tightly rolled together, rather like a leek. About ten leaves are visible at any one time; within the stem an equal number await their turn to develop as others die. Banana leaves will often be tattered by the wind, but without deleterious effect. Christopher Fryke, a ship's surgeon travelling in the East Indies in the seventeenth century, described the leaves of what he knew as 'bissang figs': 'The leaves are so large that one of them will shelter a man from the Sun and Rain . . . which makes some people apt to believe they were the Leaves which Adam and Eve made their Aprons of, after the Fall.' (See Plate 2.)

After producing 40–50 leaves and still less than a year old, the plant begins to flower. The flower stalk emerges from the

2. 'Indian Figgs'–bananas: plants, flowers, and fruit; lithograph (Nieuhoff, 1704).

centre of the stem, lengthens, then turns and grows downwards. At its tip are male flowers producing pollen and behind these are the female flowers which develop into bananas. Between five and fifteen individual banana 'fingers' are grouped as a 'hand' on a cushion growing out of the stem stalk. There may be fifteen hands in a bunch.

Edible bananas evolved thousands of years ago in the everwet regions; then natural crossing with a smaller, less sweet variety produced a new strain which could withstand drought, enabling it to spread into the monsoon lands with their long dry periods. Subsequently, man and Nature have conspired to produce many varieties of banana–twenty-eight varieties are grown in Thailand alone.

Wild bananas, with many seeds and little edible pulp, require pollination. Cultivated edible bananas, on the other hand, develop without pollination and have only a few or no seeds. Consequently, they must be propagated vegetatively, so that each plant is a clone of its 'mother'. Two groups of clones are particularly important: the tall Gros Michel variety with large, sweet fruits, and the Cavendish variety–shorter, disease-resistant plants with smaller fruits. Whilst two men are required to harvest a bunch of Gros Michel bananas, one alone can harvest a Dwarf Cavendish bunch. Filipino banana growers apparently have little faith in plant genetics, and have a traditional stratagem to keep banana plants short: when planting a banana, one should not look up, or it will grow tall. Similarly, it is said that corms should be planted by a bare-chested man to ensure thin peel, and that if he plants the corms when the tide is low and many stones are exposed on the beach, abundant harvests will be guaranteed. All this contrasts with the Filipino proverb 'A *latundan* banana will not bear the fruit of the *bungaran* banana', an Asian version of 'The leopard cannot change his spots'.

So elegantly packaged by Nature, the banana fruit is sealed away from the depredations of most insects, but some larger animals are able to remove the peel. Thick paper bags are sometimes to be seen covering bunches hanging on the tree: they are used both in controlling fertilization and in warding off flying foxes and other pests.

Bananas are similar to potatoes in food value, with approximately 100 calories per 100 g of fruit. Sweet bananas are easily digested and are a recommended carbohydrate food for children with coeliac disease. Bananas have good vitamin A content and some vitamin C. Incidentally, small local bananas are occasionally joined together in pairs at the stalk end, and a pregnant woman who eats one of these twin bananas is expected to give birth to twins! In Bali, the twin banana is believed to be useful as a charm in winning the affection of one's beloved.

Bananas are eaten fresh or prepared as sweetmeats. *Pisang goreng* is a Malaysian favourite of sliced bananas dipped in flour and fried in coconut oil. Bananas may be boiled in the skin and eaten with grated coconut meat, sugar and salt, whilst the Thai dish *gluay kai* encloses a ripe banana in a mixture of moist 'sticky rice' mixed with flour, which is then fried. ('Sticky rice' is a glutinous variety of rice often grown in the poorer parts of Thailand, such as the north-east.) Banana 'figs' are prepared by sun-drying slices of banana. 'Fig' was also the name first given to bananas when taken to Europe from the Middle East, presumably in the same sticky, brownish, sun-dried form.

Besides the fruit, other parts of the banana plant are utilized. Banana flowers are used raw in salads or cooked as a vegetable, while the heart of the stalk is used in Burmese curries; the stalk is traditionally used for stirring silk in the dye cauldrons. The leaves are often used for packaging food before cooking–for example, fish curry steamed in banana leaves (*haw mok*) is a Thai delicacy. The fruits may be prepared in many ways as a food or may be fed to livestock. The leaves may be used for thatching and packaging as well as for wrapping cigarettes and, traditionally, to make an antidote for snake bites. As a herbal remedy, bananas are recommended as a cool dressing for inflamed or blistered skin, and applied to the forehead to ease headaches. The juice of the trunk is supposed to be a remedy for baldness.

When bananas are to be eaten locally, they are left on the plant until they are fully grown. After harvesting, they are left to ripen and turn yellow in the shade. The vast majority of the world's traded bananas (sweet varieties) are plantation fruit, cut down when about two-thirds grown; they continue to ripen on the stem though they grow no larger.

The major world banana-producing regions now include

Brazil and Central America, tropical Africa, India, and South-East Asia. Most of the banana trade takes place between the American and African tropics and the USA and Europe, respectively. The main producers in South-East Asia are the Philippines, Thailand, and Indonesia, whose exports are usually destined for Japan, though the bulk is eaten locally.

PINEAPPLE (*Ananas comosus*)

South-East Asia has given many fruits and plants to other tropical areas, but has also gained crops from elsewhere. Perhaps the most important of these is the pineapple, introduced from South America. The Malay name, *nanas*, as well as the scientific name, *Ananas comosus*, is derived from the name *nana*, used by Guarani-Tupi Indians. First taken to Europe from the New World by Christopher Columbus, the pineapple was greatly valued by European society and became a symbol of good hospitality and the high status of the host. It has been grown in greenhouses as a dessert fruit and table decoration since the seventeenth century; the English name 'pineapple' derives from a similarity to pine cones perceived by early voyagers.

A field crop, the pineapple consists of a low bush of long (1 m) strap-like leaves which are thick and hard (Plate 3). The leaves grow as a rosette around a very short stem. Dark green with red mottles on the upper surface, the leaves are silvery white beneath, serrated along the edges, and sometimes spiny. From the centre of the rosette of leaves springs a thick stalk bearing the pineapple flower. The flower is made up of 100–200 small florets which develop into the fruit as they open, from the base upwards, over 20 days. The florets fuse together as they grow and the fruit is harvested 5–6 months later. Colour Plate 2 shows a Javanese woman with a harvest of pineapples. The size of the fruits in this nineteenth-century lithograph

3. Pineapples, flower and fruit, water-melon, and sweet-sop fruits; lithograph (Nieuhoff, 1704).

seems excessively small: wild pineapples are no more than 15 cm long, but these are even smaller. The thick central stalk persists through the fruit as a woody core which, if eaten, upsets the digestion badly. At the top of the core is a crown of leaves. The scaly 'shell' of the pineapple is made up of blossom cups, the calyces of the florets. The arrangement of these blossom cups is not random: they are arranged in two series of spirals, one spiralling upwards to the left at a gradual slope, the other up towards the right with a steeper slope.

A fresh pineapple may weigh about 2.2 kg, made up mainly of water and sugar; pineapples contain the vitamins A and B. Pineapples are occasionally sliced from crown to base rather than across, because of a slight variation in sweetness from tip to base. Pineapple rice, usually served in the scooped-out pineapple shell, is an Asian delicacy and pineapple is used as a vegetable in 'sweet and sour' meat dishes. *Nanas goreng*–pineapple slices fried in batter and served with cinammon sugar–is a popular Indonesian dessert. Other foods from the pineapple industry are sugar-syrup, which may be processed to give alcohol, pineapple vinegar, and citric acid; pineapple jam is exported.

Thailand and Malaysia are the main producers in South-East Asia. Much of the fruit is eaten locally. Of the two countries, Malaysia has the larger canning and export industry, based on the variety 'Singapore Spanish'. This industry seems to have sprung initially from the efforts of a single plantation and cannery owner, one J. Nicholson, who sent canned pineapples as an exhibit to the Colonial and Indian Exhibition held in London in 1886. The market he created rapidly outstripped supply.

Pineapple leaves may be exploited for the long, strong fibres they contain; raincoats were reportedly made from pineapple leaves in Sulawesi in the late seventeenth century. In the Philippines, the fibres are woven to make *piña* cloth for the richly embroidered Barong tagalog shirt, which is the national dress of Filipino men. When pineapples are grown for textiles or rope making, the young fruits are removed soon after flowering. A protein-digesting enzyme may be obtained from pineapples and used as a meat tenderizer (see papaya–papain) but there is very little commercial production of this. Half-developed pineapple fruits contain a poison which acts as a violent purge, a vermifuge, and an abortifacient; it was once specified for the treatment of gonorrhoea.

The juice of ripe fruits is believed to aid digestion and to calm acid dyspepsia, but Nieuhoff warned that whilst 'the Ananas is one of the most delicious fruits of all the Indies, [they] ought not to be eaten in too great a quantity unless you cut them into small slices and by pouring some Spanish wine on them, draw out the sharp humour'.

Light shade and rich soils give the juiciest fruits; the plant dislikes waterlogging, and indeed, is able to grow despite very strong sunlight and very dry conditions, when other plants would cease to function. Pineapples have 'tank roots'–the overlapping bases of the leaves act as reservoirs to collect water and small adventitious roots grow into these 'tanks'–there is no tap root.

The cultivated pineapple fruit fills and grows without pollination, but in its native home the pineapple is pollinated by humming-birds, which results in massive seed production (2,000–3,000 per fruit). Seeds are necessary for selective breeding, but are not wanted in dessert fruit. There are no humming-birds in South-East Asia, but some bats and insects can also act as pollinators, so the flowers may be bagged to prevent this.

The distinctive appearance of the pineapple has inspired one of Thailand's most popular designs for tableware–the spiralling blossom cups are represented in a blue and white design. Some rice bowls are also shaped to imitate the fruit. In Malaysia, pineapple juice plays a role in the manufacture of the kris or ceremonial knife, helping to wash away the substances which etch the blade. There are also reports of pineapple juice and sand being used to scrub boat-decks.

PAPAYA (*Carica papaya*)

Some tropical trees, including papaya, bear only either male or female flowers. The female papaya tree produces the fruits, but a male tree must grow close by, to pollinate its flowers.

A fruit of Central America, the papaya only reached South-East Asia in the mid-sixteenth century when it was taken to the Philippines by the Spanish. From there it spread to Melaka in the Malay Peninsula and rapidly throughout the rest of South-East

Asia. It is now well established as the pre-eminent breakfast fruit for foreign visitors, though some may find it best not to eat papaya on an empty stomach, first thing in the morning, as it may cause stomach-ache.

The papaya tree is a slim, pale grey column, knobby with leaf scars and topped with a bunchy mass of large, deeply indented leaves measuring 75 cm across. The tree grows to a height of only 10 m and does not branch unless the growing point is injured. It is relatively short-lived (25 years or so) but begins to fruit from the second year. Each tree will produce 30–150 fruits annually until age reduces its productivity.

The fruits are large, weighing up to 9 kg (though 1–3 kg is more common) and may be round, ovoid, or approximately pear-shaped (see a carved wooden papaya in Colour Plate 6). The thin, easily removed skin is green, turning yellow or orange as the fruit ripens. Papayas for dessert eating are cut down as soon as signs of ripening appear, leaving only 4–5 days for transport before they are ready for the table, so there was little international trade in the fresh fruits until recent years.

The flesh of papaya may be yellow, orange, or a reddish orange and the consistency is dense but soft–quite like butter. Inside the fruit, there is a 5-angled central cavity around which the small black seeds adhere. These are not eaten, except as a form of herbal medicine, as a vermifuge or to cause termination of pregnancy.

Green papaya fruits are used as a vegetable (but may be dangerous in pregnancy, according to Burkill). *Atchara* is a Philippine relish made of shredded green papaya, onions, peppers, condiments, and vegetables. Ripe fruits are served in chunks with a squeeze of lime juice or with salt and ground chilli pepper. Papaya jam may be made with papayas which are two-thirds ripe, plus a little ginger. The papaya is mostly water and 10 per cent of the weight is made up by sugars. It is a good source of vitamin A and also contains some vitamin C. There is some canning of papayas and the frozen pulp is used as a flavouring for yoghurt and some other foods. Papaya juice is exported canned or frozen.

But papayas have uses far beyond food. Whilst still unripe, the fruits may be tapped to obtain a milky latex, which contains papain, an enzyme which can digest protein. Latex tappers must be careful not to get this latex on their skin. The enzyme may be used as a meat tenderizer; cooks in the region use leaves and fruit when cooking meat. It is said that the latex has been used by South-East Asian poisoners, who have mixed it with other ingredients to cause great abdominal pain. A less sinister use is for the removal of warts and corns. In industry, the enzyme has many uses, ranging from chewing gum manufacture to ensuring that woollen garments do not shrink. The latex is extracted from the fruit once a week through three or four thin razor cuts into the skin of the fruit. The latex oozes out, drips into a container, and is then rapidly dried, either in the sun or in special ovens. Fruits which have been tapped are perfectly edible when ripe. The fruit of a tree may be tapped for about 3 years, after which the fruits (which grow at the top, amongst the leaves) are too high to be reached easily. Papaya leaves may be used as a substitute for soap for washing clothes.

Papaya plants are usually grown from seed (this is very easy and may be done at home, see Chapter 12); in commercial production, the male plants are thinned out, leaving only 1 male plant for 25–100 female trees. As the annual harvest from a single tree is so great, papaya trees require fertile soil, adequate moisture, and full sun. The trees are sometimes grown on irrigated land but care must always be taken to avoid waterlogging: if flooded for 2–3 days, the papaya tree will lose its leaves though it will probably recover. If the waterlogging lasts longer than this, the tree dies. Strong winds are another hazard, and the weak trunk with its spongy pith must be protected with wind-breaks in areas where high winds occur.

WATER-MELON (*Citrullus lanatus*)

The water-melon belongs to the large family of climbing plants which includes gourds, melons, gherkins, cucumbers, and loofahs. Like the rest of this family, the water-melon grows from a trailing vine equipped with tendrils for climbing, but as the fruits are very heavy (8 kg is common), water-melons grow and ripen lying on mounds, not hanging from the vine.

Water-melon fruits are usually spherical but some are lengthened to an ovoid shape. There is a hard, dark green outer rind which may be striped or mottled (Plate 3). The flesh of the fruit is crisp–the texture of an ice sorbet–sweet, and extremely juicy. Inferior varieties have a tendency to stringy flesh. Water-melons are, as their name suggests, mostly water, plus sugar and a fair vitamin A content. They are often carved as table decorations (see Chapter 10 and Colour Plate 21).

Small black or brown seeds are concentrated approximately half-way between the rind and the centre of the fruit. For serving, the flesh is sliced into triangular chunks and as many seeds as possible are removed. The seeds are edible: *kuaci* (Malay, derived from the Chinese for melon seeds) are parched for chewing by being heated strongly.

The fruits are ripe once the stem tendrils wither and the fruit gives a dull thud when tapped (4–5 months from sowing). Drive out along the roads in water-melon growing areas in the dry season and see stallholders selling the fruit to passing motorists. Slices of red and yellow melon are hung all over the stall, clearly advertising the fruit for miles. Water-melon must be eaten within 2–3 weeks of picking; it is also used to flavour water-melon ices.

This fruit is a native of tropical and subtropical Africa, but was taken to the Mediterranean basin in ancient times and cultivated widely. Although it has been grown in India for thousands of years, it did not spread to China till the tenth or eleventh century, presumably by way of South-East Asia.

3
Large Fruits

THE fruits collected together in this chapter share a common characteristic of size: they are all large, but the jackfruit is the largest of all. Of the three, the durian has the most powerful smell and the richest fruit, whilst the breadfruit grows on the most ornamental tree. All three distantly related species would be classed by botanists as primitive and may help us understand the evolution of the flowering plants. Each of the three is important in the diets and cultures of South-East Asian peoples.

DURIAN (*Durio zibethinus*)

It is hard for most Westerners to understand the place of durian in South-East Asian societies–yet perhaps champagne plays a similar role in the West. Above all other fruits, the durian is anticipated, discussed, shared, and maybe even saved for. It is an expensive fruit, and good ones are very costly indeed.

Barely known outside Asia as it becomes rancid a few days after ripening, the durian received a mixed press from early travellers. 'To eat Durian is a new sensation, worth a visit to the East to experience,' said the great naturalist, Alfred Russel Wallace; 'the more you eat of it, the less you feel inclined to stop.' Others have spoken equally strongly, but to opposite effect. The durian is very much an acquired taste, for the smell may well be daunting. If a durian is brought into a house, the smell will rapidly permeate both cellars and attic, filling them with a distinctive odour which many people find very unpleasant and which even Wallace found called to mind 'cream cheese, onion sauce, brown sherry and other incongruities'. The flavour of the fruit is said to be best when the smell is at its height–2–3 days after the fruit has dropped–but *aficionados* (one might almost say 'addicts') shrug off these inconveniences and praise the durian for its sweetness, richness of flavour (a flavour of strawberries and raspberries is tasted by some), and for its agreeable, firm custard texture.

The name 'durian' is derived from the Malay for thorns, *duri*. As with other fruits, varieties exist which offer differing qualities and tastes: 'golden pillow', 'frog', and 'gibbon' are the names of some Thai varieties. A lengthy acquaintance with the fruit may help in its appreciation: growing up with the scent occasionally wafting past might slowly accustom a future connoisseur, and many have noticed that once the fruit has been tasted, the smell ceases to seem so unpleasant. Certainly a discussion of the merits of a particular variety resembles nothing so much as the deliberations of a wine-tasting society. Only a connoisseur, for example, can ascertain the ripeness of the inner flesh of an unopened fruit: he does this by scraping the outer skin with a fingernail and listening carefully, his ear to the fruit, to hear whether the pulp has shrunk away from the pith.

The durian is an enigma at first sight. Pale olive-green even when ripe, with a thick shell heavily armoured with stout sharp spikes, the fruit seems to have something to hide (Colour Plate 3). On the tree, the fruit (2–3 kg in weight) hangs on a stout stalk 2 cm thick from stronger branches or from the main trunk itself. If the stalk is separated from the fruit at harvest, lifting a durian becomes a difficult problem. Whilst far fewer fruits are produced than, say, by a papaya tree, each may command a price fifteen or twenty times that of a pineapple (Plate 4).

Very faint 'seams' may be traced from the base to the apex of the fruit–the spines arch slightly over these lines which mark where the carpels join and where the fruit may be most easily opened with a sharp knife. When ripe, a durian may split open–this is not a defect–and inside will be revealed a white,

4. Durians for sale (Ben Piper).

slightly fibrous pith in which four or five large, cream-yellow segments are embedded. Each segment contains between one and seven seeds. It is this less strongly smelling, yellow pulp that is eaten. It is said that the fewer the seeds, the better the taste. In his account of nineteenth-century Siam, Bowring mentions that the Siamese passion for gambling extended to gambling on the number of seeds in a durian. Bowring also quotes from the *Itinerario* of Linschoten, an early traveller in the East. In this book, first published in 1596, Linschoten noted an odd affinity between durian and the leaves of betel-pepper: the presence of only a few betel leaves in a ship or warehouse full of durians was reported to cause the durians to spoil. Moreover, 'If one eats too much durian the maw becomes inflamed. One leafe of bettele to the contrairie, being laid colde upon the hart, will presently cease the inflammation of the maw.' Bowring concludes ominously, 'Men can never be satisfied with durians.'

The durian tree is easily recognizable in small farmers' plots or in orchards: it has a severely regular shape, with horizontal branches jutting out of a strong central trunk which continues to the top of the tree, 30–40 m high (Figure 1). The leaves are a bronze green, used in hot baths to treat jaundice in local medicines; the leaves and root together are used to treat fevers. Durian flowers, around 7.5 cm wide and coloured from white to pink, smell of sour milk. They are very short-lived: they open in mid-afternoon to be pollinated by bees and stay open all night to be pollinated by bats. The flower then falls before the next dawn. Despite their size, durians grow and ripen in only 3–4 months. In years of unusual climatic conditions in Malaysia, very abundant durian harvests may occur. Commenting on this, the German naturalist Rumphius (1628–1702) quoted the local belief that much illness follows a heavy durian crop.

The durian, tree and fruit, inspired J. Corner's 'Durian Theory' of the evolution of flowering plants in the tropical forest. This theory emphasizes the interdependence of animals and fruits, with plants providing food and animals providing the means of seed dispersal. Durian seeds, left behind on the forest floor once the forest animals have eaten the pulp, will regenerate and grow close to the 'mother tree', replacing this when the old tree dies. It is now thought that *Durio zibethinus* is the result of years of breeding by man rather than being a natural species, as it is not found wild in its place of origin, the Malaysian rainforest. The durian needs the hot, wet rainforest environment–the monsoon areas of Thailand, with long dry spells, are at the limit of its range.

Man is not alone in enjoying the durian, and whatever its other merits as a fruit, it is certainly nutritious, with large quantities of vitamins B, E, and C, as well as a high iron content. It is said that, in the forest, many mammals will wait for a durian to ripen and fall, elephants in particular. In the

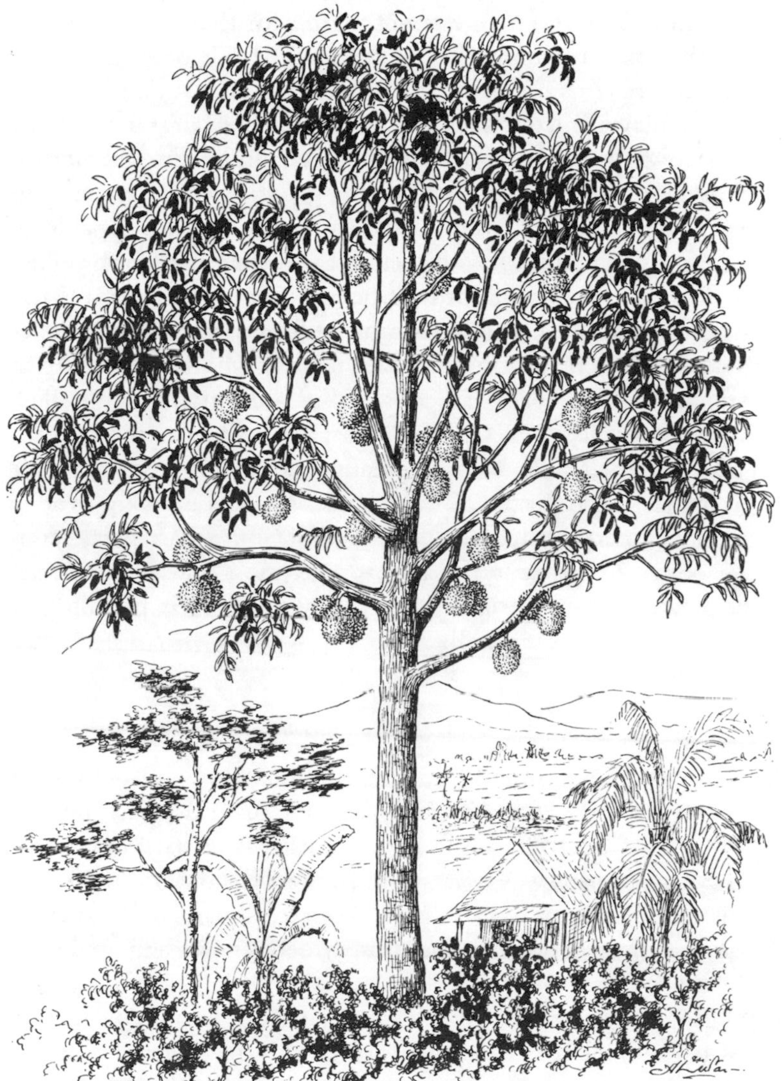

Figure 1. Durian tree (Veevers-Carter, 1984).

Malaysian forest, forest farmers sometimes build huts under a durian tree in order to reserve the fruit for themselves. Almost all the fruit reaching the markets will have been grown in durian orchards on fertile, well-drained soils. Durians are kept under surveillance as they ripen; the 'rustling' of durians is not unknown, though the fruit should remain on the tree until it falls spontaneously, crashing to the ground. Wallace mentions that falling durians were seen as a serious hazard, capable of causing severe injury with their great weight and sharp spikes.

Only the rich can afford the best durians, but the fruit is also eaten sliced and served in coconut milk, as a purée used for flavouring ice-cream, or as a paste, made by boiling the pulp in sugar. The large brown seeds are also eaten, after being boiled, dried, and fried. The Chinese extract an acid from the fruit for washing their hair. Not surprisingly, the fruit is said to have aphrodisiac properties. Despite this, it seems unlikely ever to find a wider market, for even within South-East Asia, regulations cover its transport and storage: most hotels and airlines stipulate that no durian is to be allowed on their premises, so all-pervasive is the smell.

JACKFRUIT (*Artocarpus heterophyllus*)

The jackfruit and the breadfruit are two closely related species native to Asia. The jackfruit is a native plant of India, but has been cultivated for centuries in South-East Asia. Pliny knew the fruit, which was traded from India to the African coast in his time. Jackfruit has also been taken to all parts of the tropics, but is still most popular and most widely grown and used in its home region.

Both *Artocarpus* fruits are large; jackfruit has the distinction of being the largest cultivated fruit, with the shape of an enormously overgrown pear or even, perhaps, a small barrel, measuring as much as 90 cm long and weighing up to 44 kg, though 10–20 kg is more common (Plate 5 shows jackfruit in a Thai market). The name 'jack' is believed to be a Portuguese corruption of a Malay word meaning 'round'. The thick rind is a pale green, dotted with sharp hexagonal spines: the jack, like breadfruit and pineapple, is a compound fruit, made up from the fusion of many thousands of florets on a flowering spike as they expand and grow after pollination. Inside the rind is a layer of white pith, but most of the centre is occupied by large brown seeds, each surrounded by an envelope of golden yellow flesh about half a centimetre thick. It is this yellow pulp that is eaten, either fresh or preserved in syrup. Between the individual segments are pale yellow, fleshy fibres, the remains of unfertilized florets. The flesh of the jackfruit is pleasantly firm, almost rubbery; the taste is sweet but unremarkable–'mousy' is Purseglove's description–sometimes leaving a slightly bitter aftertaste.

5. Jackfruits in the market (Ben Piper).

Although generally considered not to be a fruit for Western palates, in South-East Asia jackfruit is a popular and valuable carbohydrate food source. The seed, which is served roasted, also has plenty of carbohydrate as well as a fairly high protein content. It is said to taste much like chestnut and aphrodisiac properties are claimed for it. In traditional Thai culture, these seeds are believed to be effective as talismans, preventing guns from exploding near the wearer, or sharp objects from wounding him. The basis of this is their coppery colour–copper is a magical metal in Thai folklore. Bezoar-stones, used in faith-healing in Malaysia, are obtained from the tree; the 'stones' are lumps of hardened resin collected in cavities caused by injury to the tree. (Bezoar-stones, sometimes mineral concretions, are also found in other trees and plants. The Malay name is *buntat*.) In Malaysian herbal medicines, jackfruit leaf ash is mixed with coconut oil and applied to wounds and ulcers to speed healing, whilst the root is used to treat medical problems as different as diarrhoea, skin diseases, fever, and intestinal worms. In addition, the sap of the tree is used in treating snake bites.

When still unripe, jackfruit may be used as a vegetable or in soups. The rind is fed to livestock. Jackfruit leaves are also boiled and given to nursing mothers to increase their milk supply.

Jackfruit trees grow to a height of about 20 m and will produce fruit after only 3 years (in comparison, it takes durians 7 years to fruit). Once mature, a very productive tree may give up to 250 fruits each year: each huge fruit hangs by a stout stalk from a strong branch or the trunk itself (see de la Loubère's drawing, Plate 1, and the more accurate Figure 2). The tree provides other products besides fruit. Jackfruit timber is strong and valuable; in parts of South-East Asia where teak does not grow, jack timber is favoured for house-building because of its fine golden colour. Jackwood is also used for cabinet-making

Figure 2. Jackfruit tree (Veevers-Carter, 1984).

and for musical instruments. The yellow colour may be extracted to be used as a dye–in the nineteenth century, this dye was a considerable item of trade in Siam, where it was used, for example, to dye the robes of Buddhist monks to their shade of saffron yellow. In addition to this, a latex which may be extracted from the trunk, acts as an adhesive strong enough to bond broken crockery. As latex is also present in leaves and fruit, it is recommended practice to rub oil into the hands and forearms before cutting out the yellow pulp, for the very sticky gum cannot be removed with soap and water. This latex is the root of a Malay proverb which has been translated as 'one person gets the jackfruit but another gets be-gummed'– an Eastern equivalent, perhaps, of 'It's the rich that get the money, the poor that get the blame'.

In a rare example of South-East Asian totemism, the people of a Balinese caste claim descent from a jackfruit tree. In the beginning, it is said, the Supreme Lord made four girls and four men who had many children: 117 boys and 118 girls. The girl remaining without a husband went into the forest broken-hearted. There she found a jackfruit stump carved by Siwa in the likeness of a man. The God of Love, Semara, gave life to the figure so that she would have a husband, and the couple gave rise to the *ngatewél* clan.

BREADFRUIT (*Artocarpus altilis*)

Breadfruit originated in Polynesia and has spread from there to South-East Asia and to the West Indies. Cultivated for centuries by the Polynesians for whom it is still an important food source, the breadfruit does not usually contain seeds, and is usually propagated from root cuttings. The fruit is round or oval, about 10–15 cm long, green and studded with markings which are the traces of the flowers from which this compound fruit has developed (Plate 6). A seeded variety of breadfruit,

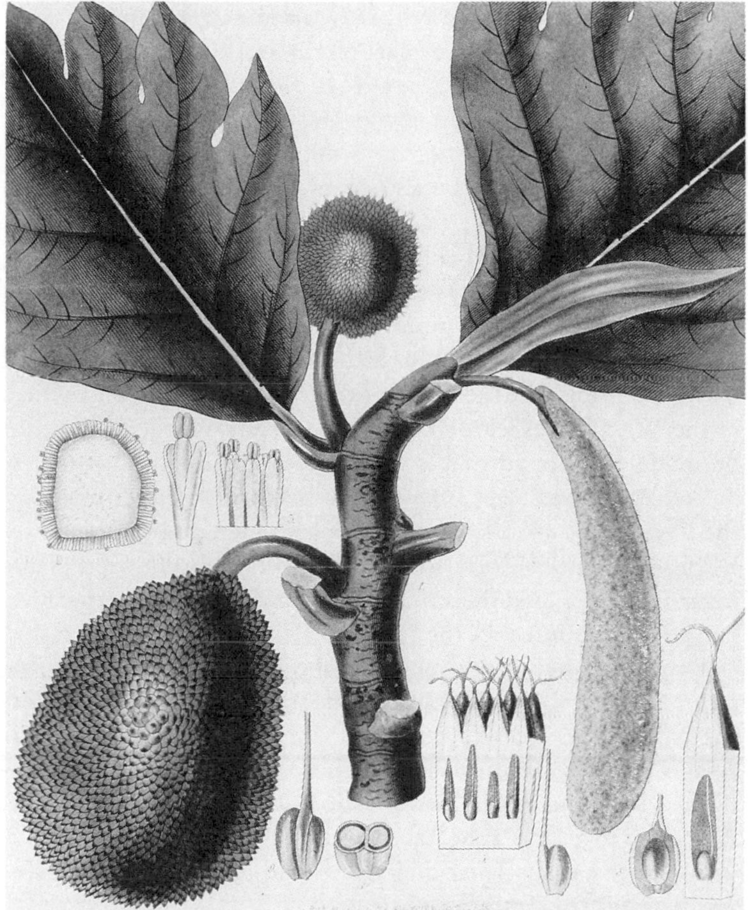

6. Breadfruit: fruit and leaves; lithograph dated 1898 (Royal Botanic Gardens, Kew).

known as breadnut, is grown for its chestnut-flavoured seeds which are served roasted or boiled. When boiled in sugar, breadfruit seeds are said to be a good substitute for marrons glacés.

The flesh is moist, pale yellow or almost white, and is a solid mass consisting mainly of carbohydrate; boiled or baked, it resembles the crumb of a new loaf. Sliced breadfruit may be cooked in a sauce of palm-sugar and coconut milk for a tasty dessert, but the fruit is more commonly used as a vegetable in curries. The flesh may be also dried and ground into flour to make biscuits.

The usefulness of breadfruit was known to early travellers to Polynesia, including Captain James Cook, who mentioned it in accounts of their journeys. One such account claims that cooked breadfruit 'is hardly distinguishable from an excellent batter pudding'. In the eighteenth century, plantation owners in the West Indies learned of this tree from which bread could be made and hoped that it might be a suitable food for their slaves. An expedition, financed by the British government of the day, staffed by a botanist and a gardener, and captained by Bligh, was sent to Polynesia in the *Bounty* to bring back planting stock, but ended disastrously in mutiny. A second expedition had to be mounted in the *Providence* to accomplish the task.

Breadfruit became popular in the West Indies. The plantation owners were lucky in that the tree they so desired prefers the habitat they had to offer: not only tropical, but an archipelago like the breadfruit's native home, well watered throughout the year and experiencing little fluctuation in temperature. As befits an island tree, the breadfruit offers a maritime by-product: the white latex which is found in all parts of the tree can be used for caulking boats.

Whilst breadfruit may not have been widely successful as a fruit tree outside South-East Asia, it will always be a popular ornamental tree wherever it can be grown. Reaching a height of 20 m, the breadfruit is one of the most startling of trees, with its very large, dark green leathery leaves casting dense shade. Contrasting against these leaves is the paler green of the fruit—which stands up rather like a candle until it is full and round.

4
Seasonal Fruits

MANY South-East Asian fruit trees bear fruit in distinct seasons: subtle change in temperature and rainfall and even day length help to concentrate the harvests of many fruit trees. Fruit set on many trees is hindered or prevented by heavy rainfall whilst a dry period will permit flower fertilization and lead to a peak of fruit production usually 3–5 months later.

The markets of South-East Asia are glutted with the greatest variety of best quality fruit in the months of May and June, before the onset of the heavy monsoon rains. In the more southerly countries, another harvest comes at the end of the year, just before the north-east monsoon. Some of the seasonal fruits–mango, mangosteen, rambutan, and litchi–are described in this chapter.

MANGO (*Mangifera indica*)

When the heat of April and May is at its height and the humidity is at its most gruelling, the mango ripens, to the delight of all.

Just like apples in the temperate countries, mangoes come in many clearly distinct varieties at slightly different times in the harvesting season, and almost all are good, in their own way. Shapes and colours differ, too, but all are a flattish oval with a 'beak' at one end; some are rounder, some are shorter, and some have a more or less pronounced beak. Colour Plate 10 shows a silver box made to represent a mango. There are mangoes which turn a pale banana-yellow streaked with green on ripening, others are pale green. Some are eaten green and hard, crisp like a cooking apple but mouth-wateringly sour-sweet; some are eaten chopped into thin slices and dipped into a mixture of chilli, sugar, and salt. However, most mangoes are probably eaten when ripe and slightly soft, when a fork cuts through the fragrant orange-yellow flesh as though through butter and the flavour is sweet, exotic, full of unexpected undertones. Inferior mangoes have a less subtle scent and taste of turpentine, while the flesh may be stringy.

The mango season in South-East Asia is lamentably brief. The main season comes before the summer monsoon and lasts only for 6–9 weeks. First a few mangoes start to trickle on to the streets and into the shops, commanding high prices. Gradually, the trickle becomes a flood and great baskets of fruits are set up at every street corner under a spreading umbrella, where a woman will sit on a low stool, waiting for customers, a reassuring sight in a fast-changing world.

Do not judge a man on face value, warns a Malay proverb: 'Horse mangoes have an ugly rind, but taste good.'

Mango flowers appear on the medium-sized, evergreen trees in December or January and the fruits gradually grow to maturity over 4 months. Rain during the flowering season will severely reduce fruit set, so fruit production is concentrated in areas with a long dry season. As harvest-time approaches, rain is needed to swell the fruit; thus, early pre-monsoon rains are called 'mango showers'. At about this time new leaves are put out, long and pendant, a translucent fresh light green against the darker background of the old leaves. To retain fruit quality, mango trees are usually grown by vegetative propagation (see Chapter 12). Young trees grow best if given some shade, but mature trees love the full sun.

The usefulness of the mango does not depend solely upon its exquisite fruit. The trunk is used, as are the leaves; even the fruit peel and seeds find a use. Mango wood is strong and resists borer attack, so the trunk is used for boat-building and particularly for dug-out canoes. Mango leaves may be fed to cattle but only when other forage is scarce as cattle fed on mango leaves will eventually become ill. The urine of mango-fed cattle is used as a dye in India, the original home of the fruit. Mango leaf ash is a popular remedy for burns and scalds. Mango seeds are believed to act as a vermifuge and are eaten in time of famine; they may also be ground to make flour.

But it is the fruit that has most applications. Apart from being eaten sliced as a dessert fruit (for example, with 'sticky-rice' and coconut milk, known as *kao niow ma-muang* in Thailand), the mango may be preserved and used in other ways. As the fruit deteriorates quite rapidly once it has ripened, mangoes are canned and processed–fortunately without losing their flavour. Traditional means of preservation include chutneys and pickles, slicing and drying in the sun, as *amchur* seasoned with turmeric, ground to a powder and added to soups and chutneys. Modern products include juice, purée, a spray-dried pulp to be used in making baby foods, mango custard, cereal flakes, and fruit bars. The processing industry makes use of its own by-products: the peel gives pectin which is used to set jellies and jams, and wine can be made from mango waste.

And mangoes are good for you–containing good quantities of vitamin A, variable amounts of vitamin C, and 16 per cent sugar. Mangoes hang from the tree on a stout stalk about 20 cm long; at least 1 cm of this is left on the fruit at harvest, to prevent the skin discolouring. When buying fresh mangoes, check the stem end for that evocative mango aroma–if the fruit is fragrant, then it has been properly ripened on the tree; but if not, then the fruit was picked while still unripe, then subsequently ripened artificially in a controlled environment.

Mango trees are very common in gardens–again, like apples in the West–because the people of the region like to have their own supply. The Thai name is *ma-muang*, and Thai house-owners favour all plants that start with *ma* in their gardens, believing this prefix–and, therefore, the tree–brings luck.

CASHEW (*Anarcardium occidentale*)

Closely related to the mango is the cashew tree, which produces both an excellent nut, widely used in cooking and in confectionery, and also the 'cashew apple', less widely known, but a useful by-product of the nut industry. (For purists, the cashew is not a nut, but a seed.) Cashew trees originated in tropical America and were brought by the Portuguese to the Malay Peninsula in the sixteenth century. Unlike the very fastidious mango, cashew is often planted on poor, stony hillsides, where little else can grow. The cashew tree may become tall and spreading; tiny pink flowers precede the fruit.

The cashew fruit is rather odd: whereas other fruit trees offer an edible fruit as an incentive to mammals to disperse the seeds which are hidden inside, cashew seeds protrude below the 'apple', which is in fact the swollen stem; Colour Plate 13 shows a cashew apple and nut, carved from ivory. After harvesting, the cashew 'apples' and nuts are separated; the juicy but rather astringent 'apple' is used in combination with other fruits in jam-making, while the nuts are shelled and then roasted or boiled. Cashew nuts contain a poison which is removed by strong heat. The fumes that rise during heating are a strong irritant and must be kept away from the face. Cashew shells contain a drying oil used in waterproofing, and as a preservative for typewriter platens. A gum obtained from the tree is obnoxious to insects and so is recommended for bookbinding; ink is made from the sap of the trunk.

All parts of the fruit are used in herbal medicines: oil from the shell of the nut is used to remove warts and corns and as an anaesthetic for leprosy sufferers; it is also considered to be an antidote to irritant poisons. In addition, the oil is painted on to wooden house posts, to deter termites.

MANGOSTEEN (*Garcinia mangostana*)

The mangosteen is one of South-East Asia's most universally acceptable fruits and is now beginning to reach European markets. Mangosteens appeal both to the palate and to the eye. On the outside, the rind of the fruit is dark purple, often marked with a yellow resin. It has small, hard, grey-green leaves at the top near the stalk, and is not especially attractive until the rind is opened by cutting around the equator of the fruit. This reveals the delicate, pearly white segments inside—a singular contrast of thick purple rind and white flesh (Colour Plate 4 and see also Plate 7). The flavour of the mangosteen is

7. A mangosteen-shaped silver box from Kampuchea (Sylvia Fraser-Lu).

extremely pleasant: very sweet, slightly acid but also fruity (a similarity to strawberries, grapes, or peaches has been suggested). The texture is soft and melting; the fruit is very juicy, with a pleasant aroma. 'The Manges Tanges', wrote Christopher Fryke in the seventeenth century, 'contains four kernels, which melt like butter on the tonge and of so fine and refreshing a taste that I have never met with any fruit comparable to it. It is generally served up at the greatest Tables as the most delicious dish that can be made; drest with Sugar and Spice and put into fine China Dishes.' Incidentally, the number of segments inside the rind can be predicted by the number of small brownish-grey triangles at one end of the fruit. Nieuhoff noticed this in the seventeenth century: 'On the top of the apple is a kind of coronet which opens as soon as it begins to ripen. The several points of the coronet have so many marks to direct you how many kernels are contained in the apple . . . those which have the most kernels are generally the best.'

Unfortunately for such a delicious fruit, the mangosteen is hard to raise: many seedlings die and it takes 8–15 years before a tree bears fruit in normal conditions. Fryke rather over-stated the mangosteen's slow propagation: 'For when you have planted it you must never expect to see it bear and if it doth chance to bear any fruit within the life of him that planted it, it comes to nothing, but withers away.' Mangosteen also requires good, fertile soils. Probably because it is rather exacting in its requirements, the mangosteen has not spread widely away from South-East Asia (the Malaysian rainforest is its native home) though it was taken to the West Indies in the mid-nineteenth century.

Mangosteen will keep for a few weeks, or a month if stored at cool temperatures. It is said that Queen Victoria offered a prize to anyone who could discover a means of transporting mangosteens to England without deterioration. Methods of extending its life are few as the delicate flavour does not survive processing; the preserve *halwa mangiss*, for example, tastes good but lacks that distinctive flavour. The same is true of mangosteen juices and jellies. Frozen mangosteens have been exported to Japan, but must be boiled before eating; it seems unlikely that the flavour would survive such treatment. Chilled mangosteen served with sherbet or ice-cream is a popular dish, and the refreshing quality of mangosteen makes it a very suitable accompaniment for durian, which is known as a 'blood heating' fruit.

The purple mangosteen rind is rich in tannin and was traditionally used as an 'astringent' to treat dysentery, as well as being used as a mordant in textile dyeing to fix the colour black.

RAMBUTAN (*Nephelium lappaceum*)

'Hairy cherry' is how visitors to the tropics sometimes describe the rambutan. There is a folk-tale told in Thailand (said to be from the pen of a king) of how a young girl chooses for her husband a man in a fearful mask topped with curly wig, knowing that under the mask he would be handsome and kind–this story refers to the rambutan's delicious flesh beneath an unpromising exterior.

The common name for the fruit comes from a Malay word for hair (*rambut*) whilst the Thai name (*ngoh*) is also the name of a curly-haired tribe from southern Thailand. The fruit is about 5 cm long with soft, fleshy spikes up to 2 or 3 cm long, all over the surface (Colour Plate 5). The easily removed peel turns from green to yellow and red as the fruit ripens. The delicious flesh is white and pearly, firm but juicy. Inside is a single large seed with a bland, nutty flavour; unfortunately, the thin seed-case sticks to the flesh. Ladies of the Thai court were once well practised in the art of removing all traces of the seed with a sharp curved knife, then replacing the fruit in its peel as though untouched. Rambutan are canned for export without seeds, and sometimes stuffed with pineapple chunks.

Rambutan grow in large bunches on a tree 20 m high, which spreads a very wide crown. The tree produces its fruit in two flushes, the main crop being harvested from June to September, with a second crop around Christmas. Like man, birds and bats are very partial to rambutan.

The rambutan originated in the tropical lowlands of Malaysia and has not been widely spread beyond South-East Asia, although its flavour is excellent and a young tree can bear fruit within 2 years of planting.

STARFRUIT (*Averrhoa carambola*)

This plant's scientific name, *Averrhoa*, is derived from that of Averroës, the renowned Arab philosopher and physician born in Spain in the twelfth century.

Starfruit has a strong, distinct aroma and is not appreciated by all. It skin is tough whilst the flesh is rather soft. Despite its shortcomings, starfruit makes an attractive garnish, sliced across to reveal a five-angled star (see the carved wooden starfruit in Colour Plate 6). Starfruit can also be made into pickles or a tasty jam–the seeds should be left in to improve the flavour of the jam.

Starfruit grows wild in the forests of Indonesia, and has spread throughout the tropics. It is a common kampong tree in Malaysia, small and umbrella-shaped, and grows well even on the coastal tin tailings. The fruit grows in groups of three or four, after the clusters of deep pink or lilac flowers have fallen. Size varies from quite small (5 cm long) to large (17 cm). Starfruit is green when unripe, turning yellow or almost orange and slightly translucent or waxy when ripe. Sour and sweet varieties exist and the fruits are rich in vitamin C and iron but lack calcium. They are said to increase the milk-flow of nursing mothers. The crushed leaves and shoots are made into a lotion to soothe chicken-pox, whilst the Chinese and Vietnamese use starfruit as an eye-salve; high blood pressure can be ameliorated by starfruit. In some parts of South-East Asia, starfruit are said to cause hiccoughs; in others, to cure it! Starfruit is also used to clean brassware.

'This tree is among the trees what the sheep are among the beasts, for they not only rob it of its flowers and fruit but also of its leaves and rinds, sometimes to the very root, as having their peculiar use in physick,' remarked Nieuhoff in 1662, commenting on the use of starfruit by the Malays and Javanese.

LITCHI (*Euphoria* spp.)

The litchi is well known as it is available, from a can, wherever in the world there is a Chinese restaurant. It is a native of southern China and evolved in a cooler climate than the tropics; in South-East Asia it is usually grown in highland areas such as the northern mountains of Thailand. Elsewhere, the tree will not fruit, as it needs a cold spell to stimulate flowering.

The litchi tree is small (up to about 10 m) and has dense evergreen foliage and small greenish-yellow flowers in large clusters. The litchi is closely related to the rambutan and the fruits have many similarities, though the texture and fragrance of the edible flesh of the two fruits is quite distinct. The red peel is marked with a pattern of small 'tubercles', or incipient spikes (Colour Plate 7); after ripening, the peel becomes brittle and eventually turns brown. The succulent edible pulp inside is a translucent white and sweetly acid, with a memorable but subtle flavour and a rose-like perfume. Bowring noted that in nineteenth-century Siam, a network of bamboo was built around every bunch of litchis, to protect the developing fruit from birds and other animals. The flesh encloses an egg-shaped seed, brown and shiny like a horse-chestnut.

Cultivated in China for 4,000 years, litchi spread through South-East Asia into Burma in the seventeenth century, and then in the eighteenth century into India, which is now the world's second largest producer. It was later taken to the West Indies and Australia.

Litchi trees are usually propagated by air-layering (see Chapter 12, Figure 3). As the tree thrives best when a symbiotic fungus (a mycorrhiza) is present, fungus-containing soil is used for the earth ball into which the roots grow from the treated twig. Trees propagated in this way will fruit after 4 years, whereas those grown from seed take 8–9 years, and may not do so at all.

The litchi has a high sugar content and plenty of vitamin C–an average helping of litchi provides a sufficient daily dose of vitamin C for an adult. Fresh litchis must be transported rapidly to market as they deteriorate after about 4 days. Litchis are also used to flavour jellies, sherbets, and ice-cream, as well as being dried and sold as 'litchi nuts', shrivelled and brown but with a pleasant flavour and a chewy, raisin-like texture. In China, litchis are fermented to make both medicine and wine.

1. A present-day market stall in South-East Asia during the main harvest season, offering: (top) melons, litchis, oranges, pears, and apples; (clockwise from left) mangosteens, tamarind pods, tangerines, rambutans, waterapples, langsat, litchis, and green mangoes (Ben Piper).

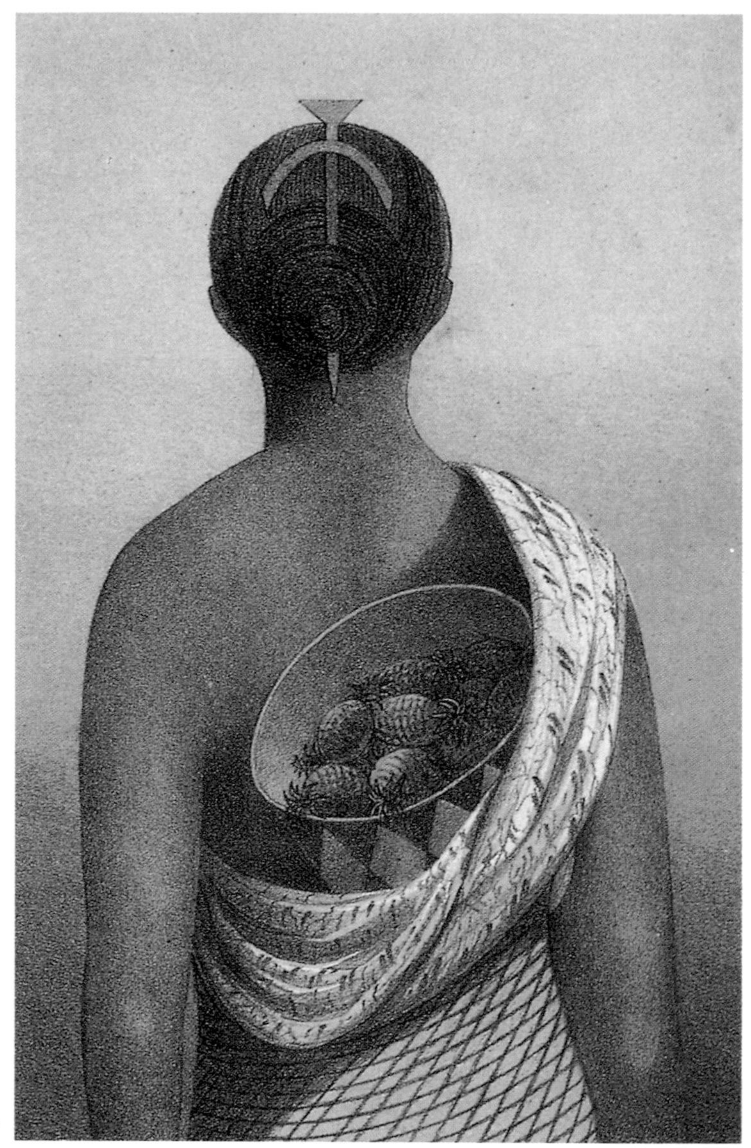

2. 'A Woman of Cheribon' with a basket of pineapples; lithograph ('Bengal Civilian', 1853).

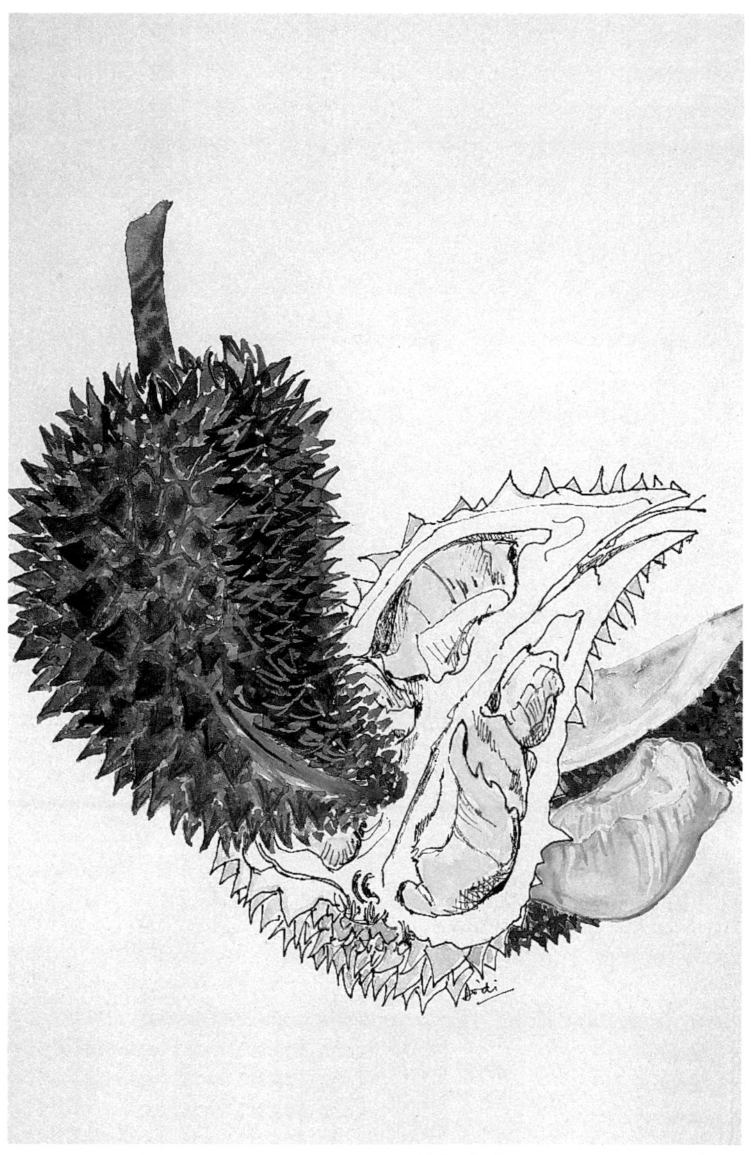

3. Durian fruit, split open to reveal the edible flesh; water-colour (Didi).

4. Mangosteen fruits and flowers; engraving (Barrow, 1806).

5. Rambutan fruits and flowers; engraving (Barrow, 1806).

6. Carved wooden fruit from Bali, including waterapple, zalacca, starfruit, papaya, and mangosteen.

7. Litchi fruits.

8. Piled offerings from a Balinese ceremony, including tangerines, apples, and waterapples (Ramseyer, 1977).

9. 'A Javan of the lower class'; lithograph (Raffles, 1817). The motif of the cloth may represent lontar fruits.

10. A mango-shaped box in silver (Sylvia Fraser-Lu).

11. Various fruits in silver filigree: mangosteen, pineapple, and (possibly) sapodilla and langsat (Vinmanmek Museum).

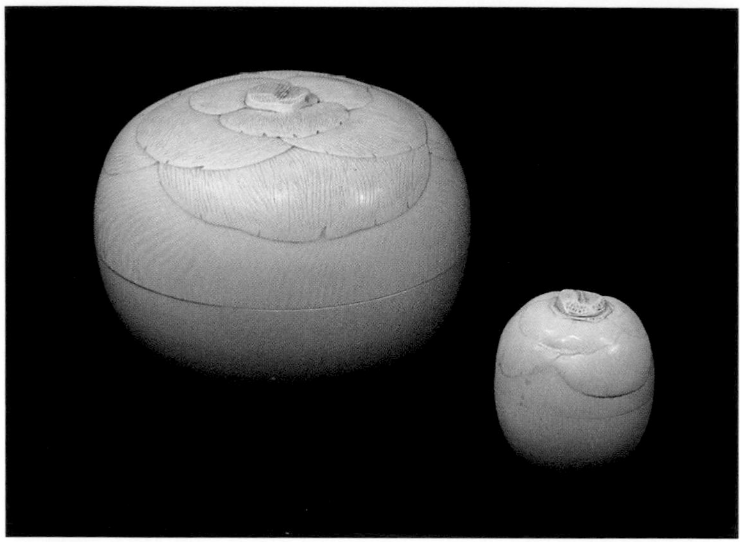

12. Lontars, fruits of the palmyra palm, carved in ivory; height of large box: 10.5 cm (Vinmanmek Museum).

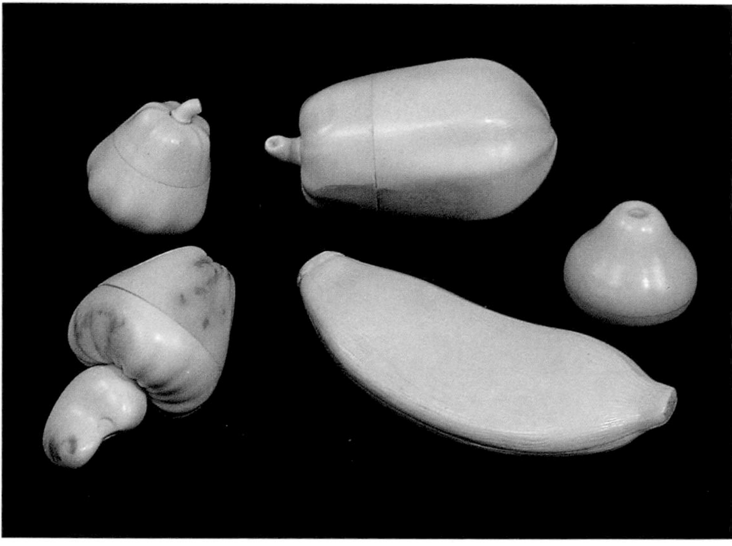

13. Several fruits carved in ivory, including banana (length: 8 cm), papaya, and cashew (Vinmanmek Museum).

14. A set of betel-leaf and areca-nut boxes of tortoise shell with diamond-studded initials of HM King Chulalongkorn of Thailand; diam. of large box: 14.5 cm (Vinmanmek Museum).

15. An areca nut, cleft in two.

16. Tamarind fruits.

17. A Burmese *kamawa-sa* religious manuscript with tamarind seed writing (Sylvia Fraser-Lu).

18. Sweet-sop fruit, leaves, and seeds; water-colour (Didi).

19. 'The Nutmeg Just Before It Drops'; lithograph ('Bengal Civilian', 1853).

20. 'Pisang Bali', a Central Javanese batik design representing banana leaves and flowers (Sylvia Fraser-Lu).

21. A carved water-melon.

22. Balinese women with ceremonial offerings (Brian Morris).

5
Citrus Fruits

CITRUS fruits are well known in all regions and are grown in great orchards across both the tropics and the subtropics, to be eaten fresh, to be canned in segments in syrup, and to be drunk as juice. The family is thought to have first evolved in South-East Asia, probably in the drier and cooler areas: most citrus species are tolerant of dry spells, but hate waterlogging, and many species can survive a light frost. Man has cultivated and bred some citrus fruits since ancient times and this work continues to the present day with crosses to produce such fruits as the 'ugli' (a grapefruit and tangerine hybrid).

Oranges and lemons are probably the best-known members of the citrus family. Though native to South-East Asia, they are now grown in the greatest quantities far away in southern Europe, south-eastern Australia, California, and South Africa—all places with long, hot, dry summers and short, cool winters. Orange-skinned oranges and yellow lemons are rarely found in South-East Asia: the colour change that occurs with ripening in drier climates is less marked here.

South-East Asia's most common citrus fruits are the limes, a sweet orange with thin peel and a less juicy orange with thick peel, and the pomelo. One or two other family members are also important in cooking. All these species have some characteristics in common. Most citrus trees are fairly small, with shiny, dark green leaves that are easily identified by the 'wings' along the leaf stalk; some species have thorns and all have pretty, sweet-scented flowers as they are bee-pollinated.

The fruits of the citrus family are grown for their juice, sweet or sharp, contained in pulp-cells growing out of the inside of the segment walls. The segments are set in a white pith rich in sugars, pectin, and vitamin C; this is not eaten but is used in jam-making to improve setting. The outer surface of the peel–the flavedo–is green and actively photosynthetic, like a leaf, in young fruits; it contains oil glands from which essential oils may be extracted.

LIME FRUITS (*Citrus aurantifolia*)

Too sharp to be eaten as a dessert fruit, the bright green common lime is an attractive garnish for rice and noodle dishes. Lime cordials are made at home and mixed with either water or soda water–salt is often added as well as sugar, to make a rehydrating drink. Lime juice also has a place in many ritual washing ceremonies, for example in wedding ceremonies in Malaysia. Lime peel is thin and adheres to the segments; it may turn yellow when fully ripe. Lime oil is extracted and used by the food-processing industry; citric acid is also made from limes.

8. Lime fruits.

Common lime is the smallest of the citrus fruits, 4–6 cm in diameter, usually round or slightly oval (Plate 8). Limes grow on a very small tree, up to 5 m tall, much branched and thorny, with small, winged leaves.

Wild lime or kaffir lime is inedible–there is very little juice under the deep green, rough, wrinkled skin. Lime oil is extracted from the peel and is used, for example, in the manufacture of 'lime essence' shampoos. In Malaysia and Sri Lanka, the juice is used in herbal ointments whilst the peel is added to stimulating tonics. The leaves are also used to flavour curries and curry pastes, sauces and soups, together with the grated or finely chopped rind, whenever a strong lemon–lime fragrance is required.

A Burmese proverb recommends against searching for the lime before the hare is caught–that is, putting the cart before the horse. There is a Malay superstition that wild lime will drive away evil spirits.

POMELO (*Citrus grandis*)

The largest of the citrus fruits, pomelo is much appreciated in South-East Asia but is less well known elsewhere–it was first taken to Europe by voyagers in the twelfth or thirteenth century, but only as a curiosity. It was taken to the West Indies in the seventeenth century by a ship of the East India Company commanded by Captain Shaddock, so in that part of the world, the same fruit is often known as a shaddock.

The pomelo fruit may weigh a kilogram or more and is shaped like a pear or a flattened globe. The skin is green, sometimes marked with pale orange, covering a very thick, spongy pith. There are 10–12 large segments, a pale yellow or pink in colour, each surrounded by a very tough membrane which must be removed before eating–a delicate art as the segments fall apart easily as this is stripped off. For the Kampuchean salad *nheam krauch thlong*, pomelo segments are separated into individual pulp-cells and mixed with prawns, toasted coconut, and very crisp bacon. Pomelo juice is sweet, but not strongly so, and less tart than grapefruit. The fruit is a particular favourite of the Chinese in South-East Asia, and is associated with all Chinese festive occasions. Pomeloes are always placed upon the household altar of Chinese homes on the fifteenth day of the eighth moon, the Mid-autumn or Mooncake Festival, when cakes and fruits are offered to household deities.

Pomelo marmalade is made in a small way. The leaves are used in aromatic baths and an infusion made of pomelo leaves is used as a sedative for 'nervous affections' in the Philippines. In Malaysia, a lotion made from boiled pomelo leaves is applied to swellings and ulcers.

Pomelo trees are considered to be native to Thailand and Malaysia, flowering and fruiting all year round. The sweetest and juiciest pomeloes are said to grow in slightly brackish water on low ridges in Thailand, where pink varieties are especially favoured; inferior fruits are rather dry. The tree is small and spiny with large cream-petalled flowers.

A chance cross between a pomelo and a sweet orange may have given rise to the grapefruit, or grapefruit may be a variety of pomelo; the two fruits are certainly very closely related. Grapefruits first evolved in the West Indies and are rarely seen in South-East Asia.

SWEET ORANGE AND TANGERINE
(*Citrus sinensis* and *Citrus reticulata*)

The sweet oranges found in South-East Asia are often green-skinned or a pale orange streaked with green, and lacking the sharp taste of Valencia oranges. Sweet oranges are far less juicy than those grown in the Mediterranean basin, and are usually served peeled and sliced into quarters. When sliced open, we see deep green peel, brilliant white pith, and bright yellow-orange fruit inside.

Sweet oranges make significant gifts at Chinese New Year ceremonies in Malaysia and elsewhere. Two varieties are particularly common because of the meanings of their names in Cantonese. One, the *gán* variety, is synonymous with gold and this fruit is used to symbolize all that is auspicious, golden and sweet. *Jú* oranges, on the other hand, symbolize good luck, another common Chinese blessing.

Mandarin and tangerine oranges are two varieties of *Citrus reticulata* (the satsuma, grown in Japan, is another). 'Mandarin' usually refers to the fruit with yellow skins and 'tangerine' to those with deep orange skins. Tangerines are also squeezed for their juice which has a very slightly pinkish tinge and is often served with a little salt. Mandarins are thought to have evolved in the cooler parts of South-East Asia, once known as Cochin-China (now Vietnam and Laos), where cold winds from China sweep southwards during the winter months, bringing temperatures quite low. This is the hardiest of the citrus fruits and has been taken to temperate countries to be grown in orangeries and greenhouses as well as to all tropical and sub-tropical countries. These citrus fruits are included in the piled offerings shown in Colour Plate 8.

Philippine horticultural folklore has it that grated coconut spread around the roots of an orange tree will make it bear abundant fruits.

FINGERED CITRON OR BUDDHA'S HAND
(*Citrus medica* var. *sarcodactylis*)

Another citrus species with some local importance in South-East Asia is the citron. The citron was used by the Romans and made into candied peel. The *sarcodactylis* variety, the 'Buddha's hand', is quite rare; in this variety, the pale yellow carpels are incompletely fused together and remain free like fingers on a hand–there may be five or more of these 'fingers' (see Plate 9).

Fingered citron has been used in herbal preparations since ancient times; it probably evolved in south-western Asia and then spread through India towards the south-east. It is used in Buddhist observances (the peel is particularly fragrant) and may also be used as an ornamental element in fruit arrangements and temple offerings; it is not eaten as a dessert fruit.

9. A fingered citron carved from wood and bamboo; Chinese, seventeenth to eighteenth century (Victoria and Albert Museum).

6
The Myrtle Family

THE myrtle family is a huge group of trees and shrubs ranging from the great Eucalyptus genus to the common European myrtle. As well as guava and waterapple, the family includes a great many other useful plants: timber trees and spice trees in particular (see also cloves–Chapter 9).

GUAVA (*Psidium guajava*)

The small guava tree is grown throughout the tropics, in all sorts of environments. It originated in the American tropics and reached Asia in the seventeenth century.

Guava fruits, available all year round, are either round or pear-shaped, with a thin green skin which becomes yellow on ripening (Plate 10). Wild fruits are small, but cultivated varieties may attain 13 cm in diameter and weigh 700 g. Whilst still unripe, the flesh is hard and slightly gritty (like a pear) and the flavour is sweet but not strong, perhaps a little like apple; when ripe, the fruit is softer and highly aromatic. In Thailand, the fruit is eaten whilst it is still hard and unripe, as Thais dislike the smell of the ripe fruit. Much of the centre of the fruit is occupied by the extremely hard, white seeds. To avoid tooth-cracking, the fruit is cut to remove the seeds, leaving crescents of greenish-white flesh, 1–2 cm thick; pink and red varieties also exist. Mature guavas do not keep well, so in the tropics, the ripe fruit are often shipped at night by barge to prevent spoilage by sunlight or bruising.

The texture of guava and its cooking properties make this a substitute for apples in the tropics, where it may be used in pies and pastries. For export, it is puréed and canned or made into

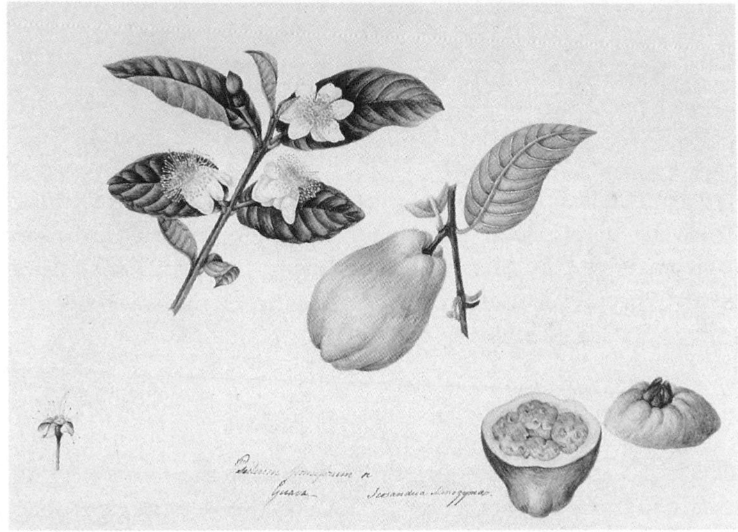

10. Guava, fruit and flowers; lithograph (Royal Botanic Gardens, Kew).

guava nectar. As the fruit has a high pectin content, it can be made into jams and jellies; it also has a surprisingly high vitamin C content and an average helping each day will provide more than enough vitamin C for an adult. The fruit is also rich in iron and calcium.

The guava tree has some other uses: in India, the leaves are used for treating wounds and as a remedy for toothache; a decoction of leaves and root is used in the Philippines as a remedy for swollen gums and diarrhoea. The bark of the pale brown, mottled trunk is also very astringent and is used for dyeing and tanning. Silk dyers in Pekan, Malaysia, dye silk black with an infusion of several tannin-containing plants heated, cooled, and reheated over a week. The plants used include guava leaves, husk of young coconut, leaves of the Chinese tallow tree, rambutan fruits, and mangosteen husks. Guava wood makes good firewood.

WATERAPPLE (*Eugenia acquaea*)

Children love the crisp, juicy flesh of waterapple, with its thin outer skin coloured from green to a ruby red; waterapples top the piled offerings shown in Colour Plate 8 and there is a carved waterapple from Bali in the basket of wooden fruit shown in Colour Plate 6. The flavour is quite variable: some waterapples have a thin, slightly acid flavour, but others have a full and fruity apple flavour–though when cooked and sweetened, the flavour is more like a pear. Waterapple is also eaten dipped into a mixture of salt, sugar, and chilli. The fruit is bell-shaped and about 5 cm across at its widest part. However, not much of the fruit is edible: the centre is filled with woolly fibres, within which brown seeds are located. Ants love to tunnel into the fruits and frequently spoil them.

Although it is slow-growing, the waterapple tree is a favourite garden ornamental because its large evergreen leaves cast deep shade. The flowers are remarkable, with cerise to deep crimson petals and tufts of pink stamens, and are sometimes eaten in salads. Medicinal properties are claimed for the fruit.

The roseapple, *Eugenia jambos*, is another myrtle fruit. Small and egg-shaped, its white flesh is delicately scented with roses. This tree is linked in myth with the golden fruit of immortality.

7
Palm Fruits

PALM trees dominate the landscapes of South-East Asia. Ornamental palms are commonly planted in parks and gardens; in the countryside, palmyra and coconut are perhaps most common. These two palms provide food products as well as many of life's other necessities, including thatch, timber, paper, and oil.

Palms are of two broad types, the 'feather' and the 'fan' palms. Once a palm is full grown, it carries a constant number of leaves, losing an old one from the bottom of the crown as a new leaf pushes its way up vertically into the sky: tightly furled before it opens, this 'sword leaf' is a fresh light green. There is a Malay proverb which, translated literally, recommends that the new palm leaf should not laugh to see the old leaf fall. In other words, those who laugh at another's misfortune may find their own fate is similar; people in glass houses, perhaps.

Palm flowers–insignificant, pale, and short-lived–are located on a many-branched flowering stalk (the rachis) which is initially protected by a strong outer case called the spathe. This outer protection falls once the fruit starts to grow.

COCONUT (*Cocos nucifera*)

Along the coasts of South-East Asia, the coconut palm finds the environment it prefers: sandy soils are well drained, temperatures do not fluctuate, and rainfall is abundant. Also, as the enduring image of the tropical paradise island shows, the blue skies provide plenty of light.

As with several other palms, the coconut flower spathe

produces a sugary liquid which becomes slightly fizzy with fermentation. If this liquid is left to ferment throughout a day, it is quite alcoholic by evening, when it is drunk as toddy. The liquid may also be boiled down to obtain a brown sugar called *gula melaka* or *gula kelapa* in Malay.

The outside of the coconut is a hard, smooth layer, which is usually green, though some varieties, such as 'king coconut', are golden or even red. Inside this is a thick layer of fibres known as 'coir' which surrounds a shell with three 'eyes'; one of these is a soft area through which the emerging seedling will push its way to grow into a palm. Inside this inner shell is a whitish, jelly-like layer which is the coconut meat as it starts to form and this encloses a milky fluid and the small, peg-like embryo of the palm tree which lies under the soft 'eye'. Coconuts are ripe for picking 12 months after the flowers are pollinated.

Well sealed inside its almost impregnable multi-layered shell, coconut water is of great purity. This purity is recognized by Nicobar islanders, who wash their babies only in water from a young coconut until the age of two months. Coconut water is used in medical laboratories as a medium on which to culture cells; it is also occasionally used as an intravenous drip: the solution, being isotonic with the body fluids, comes ready-made for the purpose. In plant-breeding laboratories, the tiny plantlets of tissue culture are sometimes nurtured with coconut milk.

When coconuts are left to ripen on the tree, the coconut water gradually disappears, forming the white coconut meat. After harvesting, this coconut meat is extracted and used in several ways. It may be grated and squeezed to give a rich coconut milk, or grated and infused in water to give larger quantities of a less rich milk. Coconut milk is used as the basis of South-East Asian curries in much the same way as stock is used for stews in the West. 'Coconut cream' is skimmed off this milk when the cream has risen to the surface if the milk is left to stand. The cream is used in curries and to decorate foods as well as in confectionery and sticky sweetmeats. The Philippine dish *bibingka* deserves a mention: ground rice, coconut milk, sugar, and eggs are cooked together over charcoal. Once set, the mixture is topped with white cheese or salted duck eggs; the dish is served with tea.

'Copra' is the name given to dried coconut meat–after the shell has been split open, the meat may be left to dry in the sun or may be gouged out and dried artificially, sometimes in kilns fuelled by coconut husks. Copra may be pressed to give coconut oil used in the manufacture of soaps, cosmetics, and even candles and baking fat. For preservation and export, grated copra is dried and sold as desiccated coconut. The clear, firm embryo in the centre of the nut is usually served in coconut milk; it has a flavour so subtle it is not always perceptible.

Coconut coir has long been used for mats and coconut shells may be burned as fuel or made into ladles and bowls. Rudimentary musical instruments can also be made by scooping out the meat from the undamaged shell, or leaving it to rot away. Very occasionally, a small, pearl-like stone may be found inside a coconut. Wallace believed that these 'coconut pearls' may arise where a germinating shoot is unable to penetrate any of the 'eyes' in the shell and so fails to emerge.

About 3,000 years ago, coconuts floated across the ocean to establish themselves in India, then gradually displaced the palmyra palm there (see pp. 53–5). Early voyagers, including Marco Polo, knew the fruit as *nux indica* (Indian nut); it is thought that the name 'coconut' springs from the Portuguese for monkey or bogey figure (*quoque*), as Portuguese travellers who brought the fruit back to Europe saw a resemblance to a monkey's face in the hairy sphere with its three 'eyes'.

Monkeys have always loved coconuts. Christopher Fryke, travelling in Java in the late seventeenth century, tells, with a fine disregard for grammar, how coconuts were used to trap monkeys. To do this, a small hole was made into the fruit, just large enough for a paw to enter. As soon as a monkey 'espied the hole, they wanting to get at the Kernel, would strive hard but, they would get their paws in: And when they go to take them out again, they have not the sense to squeeze their Claws together to slip their Paws out, as they had to get them in. Beside that, the surprize which the person causes who watches them, makes them less able to rid themselves of their Manacles; and as they went to run down, they fall with the very weight of the Nut, which is or may be five or six pounds weight.'

There are strong local traditions regarding the planting of coconuts in Malaysia–for example, coconut should be planted after a good meal. Moreover, the seed nut should be thrown into a previously prepared hole without straightening the arm–to straighten it might lead to the fruit stalk breaking, years later. Custom further advises that by planting in the evening, the fruit grower is ensuring that his palms give fruit while still near the ground. Coconut palms grow to a height of 20–30 m (see Plate 11) and have a life-span of 80–100 years, producing fruit from about 5 or 6 years but reaching peak production around the thirtieth year. During a palm's productive life 3–6 coconuts can be harvested a month, or a total of about 3,500 over 60–70 years.

Throughout its life, the palm also produces enormous leaves, each of which may be several metres long, composed of many leaflets which can rotate slightly to follow the sun–this is indeed a sun-loving plant. Once mature, the coconut carries 25–30 leaves; a new leaf appears at the top, growing from the central growing point, as an older, lower leaf falls. Each leaf can weigh 10–15 kg, so a falling leaf is to be avoided as much as falling coconut. The leaves can be used for thatching and for mats (the *peng* mat of coconut leaves is used in Malaysian séances, when spirits are summoned). The coconut palm

11. Coconut palms and fruits and a cinnamon tree; lithograph (Nieuhoff, 1704).

'cabbage'–the unopened leaves at the top of the stem–is made into Filipino *atchara*, or salad.

James Low, writing in the British settlement of Penang in the nineteenth century, remarks that coconut palms are extremely likely to be struck by lightning (killing the tree) and so they should not be planted close to a house. Nevertheless, even after death, the coconut palm is still useful: its trunk is strong and insect-resistant, and so is used in house construction.

Several parts of the coconut palm are used in South-East Asian herbal medicines: to cool fevers and sunstroke, the pulp of the young fruit is given and coconut oil is applied to the head. Coconuts are very commonly used in the rituals associated with childbirth throughout the region–some of these are described in Chapter 11. Coconut root is eaten to strengthen the gums or teased into a fibrous tuft and used as a toothbrush. Sanskrit authors recommend the flowers for the treatment of diabetes, dysentery, and even leprosy. In addition to all this, nursing mothers are encouraged to drink coconut milk, as it is said to improve their own milk production. Coconut oil is widely used as a cosmetic for skin and hair in South-East Asia, often perfumed with herbs and flowers. Balinese women massage *boreh* (a yellow paste made with macerated herbs and spices including clove, nutmeg, and turmeric) into their skin. Less direct ways of courting attention also involve coconuts: The Balinese use twin coconuts as a love charm, as well as oil from a coconut under which a pregnant woman has sat. As a more desperate measure, the unrequited Balinese lover may sit up through the night staring at the flame of a lamp made from a new coconut, burning freshly made oil. If the suitor concentrates on his beloved's face, his love will eventually be returned.

Thus the coconut palm is essential to the Asian small farmer: it gives him food and drink, builds his walls and thatches his roof, decorates his garden and fuels his cooking stove. Small wonder it is known locally as the 'tree of life' and the 'tree of heaven', and a coconut palm is often planted to celebrate the birth of a child.

LONTAR OR SEA APPLE (*Borassus flabellifer*)

Whilst coconuts grow in well-watered areas, the palmyra palm is found growing in drier, monsoon areas away from the coast, most noticeably on the raised ridges between paddy fields. Once the rains come, these paddies fill with water which shows green as the young rice shoots emerge. The tall, graceful palmyra palms are reflected across the surface of the green water, in one of the most common and peaceful images of the

Asian countryside. There is a Burmese saying: 'Fall in love with a palmyra leaf if, for you, it's a beautiful fairy'–a South-East Asian version of 'Beauty resides in the eye of the beholder'.

The fruit of the palmyra palm is the lontar–Colour Plate 12 shows a lontar carved in ivory. This fruit is less well known to visitors than the coconut, but to rural people, it is certainly as valuable as the coconut. Lontar fruits are sold in markets and have flesh similar to that of the coconut–translucent when very young, becoming whiter and harder and tasting more strongly of coconut. The edible part is enclosed within a pale brown inner shell, removed before eating. There is also some watery liquid inside the nut, which may be drunk iced.

But the palmyra palm is grown not so much for its fruits as for the coarse, caramel-flavoured sugar or 'jaggery' it gives. This is made from the sugary liquid inside the spathe or flower case before it opens. To tap this palm sap, the two parts of the spathe are tied together at the tip, then the spathe is beaten with a mallet at regular intervals over a few days. A thin slice of the spathe is removed and it is bent downwards to allow the liquid to drip out and so be collected. The liquid is then boiled down to obtain a brown sugar which is preferred by cooks in South-East Asia, who believe it to be less cloying than cane-sugar. Palm sap can also be left to ferment for a day to produce toddy. When a spathe is tapped for sugar, no 'sea apples' will be produced. Incidentally, in some parts of Indonesia, if a tapper is killed in a fall from a palmyra palm, it is customary to fell the tree and use it for his coffin. In the eastern Indonesian island of Roti, for example, custom has it that women should be buried in coffins made from female palmyras, and men in coffins from the wood of male palmyra palms.

A palmyra palm will be 10–20 years old before its first flowers appear and tapping can commence, but once mature, there may be several spathes all producing palm sap simultaneously–as much as 3 litres a day per spathe. As a full-grown palmyra palm may eventually reach 25 m, some convenient means of reaching the flowers must be found; flower-tapping is not a job for monkeys. This is why rope ladders or spikes are usually to be seen on the trunks of palmyra palm. Each palm is climbed every day over a 5-month period, usually in the early morning; the palm will continue to be tapped for up to 30 years.

Whilst the coconut is a feather palm, the palmyra has great fan leaves measuring up to 1 m across. These leaves have traditionally been used for sacred manuscripts. The text is incised with an iron stylus on to the narrow strips of leaf. Threads are inserted through the corrugations of the fan so that the leaf may be pulled together and the book 'closed'. Raffles reports that the Buju ruler of Celebes (Sulawesi) would summon lesser members of his court by sending a palm leaf with some of its points knotted together. The number of knots represented the number of days in which the man was to present himself.

ZALACCA (*Salacca edulis*)

The zalacca fruit comes from one of the group of palms which do not form trunks, but rather sprout their leaves from ground level. Cultivated for their fruit, zalacca palms are found as dense thickets, usually on low-lying, flat, wet land. In the words of one early nineteenth-century traveller, the zalacca grows 'in swampy places about Malacca, the Tenasserim Provinces and Burmah'.

The central shaft of the zalacca leaf is densely covered with spines, forming a prickly fence around the bunch of reddish-brown fruits that protrude from near ground level (Plate 12). Zalacca fruits are quite similar in form to the fruits of the rattans, their close relatives, with overlapping scales covering a round or egg-shaped surface. The taste of the zalacca is much

12. Zalacca fruits (Didi).

stronger than that of coconuts or lontars–there is even a suggestion of jackfruit taste once the flesh ripens to a sour-sweetness. Zalacca palms flower in the cold season and the fruit ripens in June and July.

BETEL-NUT (*Areca catechu*)

Outside Asia, the betel-nut is probably best known from the books of writers such as Somerset Maugham who described the Asian habit of chewing betel quids and then spitting out saliva, leaving splashes of red on buildings and footpaths.

The betel-nut or areca nut is the seed of the Areca palm (see Plate 2, centre), grown widely throughout India as well as Indonesia, the Philippines, and Malaysia–where the state of Penang takes its name from the Malay for betel-nut, *pinang*. The nut is quite large, weighing 10–20 g, and enclosed in a fibrous husk which is yellow or orange when ripe. There is no edible flesh. The fruits are sold still attached to the branching stem which was once the flower stalk.

For chewing, the betel-nut is sliced finely and shreds are combined with a small amount of slaked lime; the mixture is then wrapped in a leaf of betel pepper (*Piper betle*). Other flavourings may be added, such as clove, cardomoms, tobacco, gambier leaves, or catechu (the chopped heartwood of the tree *Acacia catechu*). Semi-medical benefits are said to result from chewing: the elimination of bad breath, wind, phlegm, and 'impurities', whilst the mouth is said to be made more beautiful–in fact it is stained red, whilst teeth turn black and are ground down by the lime over years of chewing. In local herbal medicine, areca nut is used to treat intestinal worms and to terminate pregnancy.

Betel-nuts are produced when the palm is between about 8 and 40 years old. The fruits and flower stalk are cut from the tree with a curved knife atop a long pole. The husks are removed and the nuts are processed by drying and/or smoking; the tannin content of the nuts may be reduced by boiling them before drying.

An attractive palm, areca is often grown as a garden ornamental. It is slender, topped with a crown of 8–12 feather leaves and grows to a height of about 30 m. The palm prefers tropical climates influenced by the sea and requires abundant rainfall.

Chewing betel is considered to be polite behaviour at most auspicious occasions and religious festivals and observances (see Chapter 11). The habit is shared by people of all walks of life, rich and poor, male and female. Perhaps nowadays it is becoming a habit of the middle-aged and the old: something that people expect to take up when they are older, perhaps when they have more time to prepare the quid. However, like cigarettes, chewing betel-nut is not immediately pleasurable for beginners, who may feel dizzy or feel a tightening and burning sensation in the throat. Even the experienced chewer will suffer dizziness if he chews betel on an empty stomach; this sensation gave rise to the Asian proverb, 'eating betel, suffering from vertigo', which refers to attempting a task for which one is ill prepared, and then becoming confused.

Whilst betel-nut chewing is obviously an acquired taste, anyone with an interest in regional handicrafts will be impressed by the paraphernalia that surrounds the habit: beautiful boxes of lacquerware, bamboo, silver, and wood are made with many compartments, to house the various ingredients (see, for a particularly splendid example, Colour Plate 14). Several tools are used to prepare the quid for chewing: pestle and mortar, nut-slicers, and lime spatulae are some of the finely decorated tools used.

The betel-nut is linked in the South-East Asian imagination, with love and marriage, symbolized perhaps by the inseparability of betel-nut and betel pepper. A Malay proverb uses the areca nut as a symbol of a good match of man and wife—'Like an areca nut cleft in two' (Colour Plate 15)—whilst a Vietnamese folk-tale tells of the origin and discovery of the betel-chewing habit.

Once, in the reign of the semi-mythical King Hung IV, two handsome brothers studied at the knee of a teacher with a beautiful daughter. Tan, the elder brother, fell in love with the girl and requested her hand in marriage. After the wedding, the young couple became very much wrapt up in each other and Lang, the younger brother, felt forgotten and neglected. One night, sleepless with loneliness, Lang went out into the forest and walked on until exhaustion and sorrow claimed him. He sat beside the riverbank sighing and sorrowing, until he died of grief and fasting. Seeing this, the gods changed his body into a rock of limestone, as a symbol of steadfast devotion.

A few nights later, Tan realized his brother had gone and set out to find him, acknowledging and regretting that he had grown away from Lang. Tan walked many miles until he, too, sank down upon a rock beside the murmuring river, crying for his brother and weeping till death overtook him. The gods changed Tan's body to a palm, slim and elegant on the limestone rock beside the river.

When Tan's wife realized that her husband was missing, she, too, set off in search. After walking many miles, her eyes blinded with tears, she reached the riverbank and wept till death for her lost husband. The gods changed her to a vine which clung to the palm.

A great drought came and all the land was dry and withered. Only the palm and the vine upon the the limestone rock remained green. King Hung brought his court to see this phenomenon of verdant tree and vine in the brown landscape. He ordered the nut of the palm and the leaf of the vine to be brought to him. Detecting a faint aroma, he placed both in his mouth and chewed. The king was soothed by the sweetness and fragrance of the mouthful and encouraged his courtiers to try some, too. And from such beginnings, we are told by the story, did the fame of betel-chewing grow—which is perhaps why the Vietnamese suggest that chewing betel is 'the beginning of all conversation and love'.

8
Less Common Fruits

THE five fruits described in this chapter (sapodilla, tamarind, sweet-sop, longan, and langsat) are eaten and enjoyed by the people of South-East Asia during their fruiting seasons, but are produced in rather smaller quantities than most of the other fruits in this book and, apart from the sapodilla, are not widely known outside Asia.

SAPODILLA (*Achras zapota* or *Manilkara achras*)

Sapodilla is one of several fruits which spread from Central America with the Spaniards who went to colonize the Philippines in the sixteenth century. It then spread from the Philippines to the rest of South-East Asia. In Malaysia, the fruit is *ciku*; in Thailand, *lamut*. The English and American name 'sapodilla' is derived from the Mexican name, *zapotl*.

When ripe, the sapodilla is remarkable for its extremely sweet, honey-like flavour, though while unripe, a milky sap makes it unpleasant and astringent. The flesh is light brown, becoming darker as it ripens, and has a rather sandy texture; it is carved as a decorative addition to a dish of dessert fruit. Sapodilla is invariably eaten as a fresh fruit as processing destroys its distinctive flavour, so it is not seen outside the tropics.

Sapodilla trees fruit throughout the year and yields are high: a mature tree can produce as many as 3,000 fruits in a year; they are egg-shaped and have a thin peel, rusty brown in colour. Another product of this tree is chicle, obtained by tapping the trunk, which is used in the manufacture of chewing gum.

SWEET-SOP (*Annona squamosa*)

Sweet-sop, sugar-apple, and custard-apple are the names given to a fruit of South American origin which, like the pomegranate, is packed with seeds. The edible pulp is a thin envelope surrounding each of the many seeds. The pulp is creamy white in colour and slightly granular in texture (in fact, not unlike custard with granulated sugar stirred in, unmelted). Eating a sweet-sop can be quite hard work, though rewarding—the taste is sweetly bland but distinctive. A delicious ice-cream can be made with the fruit. In Malaysia, the fruit is commonly known as *seri kaya*—'rich in grace'.

Sweet-sop is another compound fruit, produced by the fusion of many florets, like pineapple and jackfruit. Unlike these larger fruits, however, the separate parts of the sweet-sop are not tightly fused under a common rind, but will fall apart quite easily. The outside of the fruit, green even when ripe, appears rough and almost scaly as the green outer surfaces of the segments bulge against each other (Colour Plate 18). The apple-sized fruit are approximately heart-shaped and hang on a stout stalk from the branches of the small tree, not more than 5 m tall (see Plate 3).

Crushed, the very shiny, dark brown seeds of sweet-sop are sometimes mixed with coconut oil and applied as a treatment for headlice. They are extremely toxic, and can be used to terminate pregnancy.

TAMARIND (*Tamarindus indica*)

Although known as the 'dry date of India'—in Persian, *tamr* is a dry date, and *Hind* is the the River Indus—the tamarind probably originated in Africa. It reached India in ancient times and it is still most widely grown there. From India, the tamarind tree spread into South-East Asia.

Tamarind fruits are pod-shaped and look a little like butter beans; the tamarind is a member of the vast Leguminosae family (which includes peas, beans, and mimosa). When ripe, the outer case is brittle and brown, rough and scurfy on the surface and about 5–10 cm long (see Colour Plate 16). Inside the pod there are up to 10 seeds surrounded by a dry, sticky pulp. It is this pulp that is eaten and which has a taste reminiscent of a hot, unsweetened lemon drink. Filipino lore has it that, to make the tamarind fruit as sweet as sugar, sugar should be mixed into the earth in which the seed or seedling is planted.

The tamarind is considered a great delicacy in South-East Asia, with prices to match: gift-wrapped, a kilogram of perfect pods can cost almost as much as a durian. Tamarind confectionery is also available, such as tamarind balls made of tamarind flesh mixed with sugar. The pods are eaten whole, peeled and seeded, and then pickled and dipped in a chilli and sugar relish. Tamarind is also used to flavour food, in curries, chutneys, and sauces, and a drink. Fish in tamarind sauce is a Philippine speciality. High in protein, the seeds are eaten roasted, boiled, or ground into flour after the hard seed-coat has been removed; this flour is used to make vegetable gum for the food-processing industry. The flowers and leaves of the tamarind tree are also used in curries and soups.

In herbal medicines, tamarind juice is claimed to be a remedy for dysentery, whilst a decoction of the leaves is used to treat worms in children. In Kampuchea, the juice of young leaves, heated and filtered, is used in eyedrops to treat conjunctivitis. Tamarind fruit left past the point of ripening is sometimes used to clean copper and brass, as the overripe fruit contains tartaric acid, which will also rot cloth. Tamarind ink is made in Java by burning the bark of the tamarind tree.

The tamarind tree is much favoured as an ornamental: its delicate feathery foliage trembles in the breeze, contrasting with a trunk like cast iron, almost black and deeply fissured, with jutting branches. The tree resists urban pollution and is frequently used as a shade tree to line important avenues or planted in Asia's verdant parks and gardens–this is a far cry from its native home in the drier savannas of Africa, amid the thorny woodlands of acacia and baobab.

The tamarind tree is a favourite of the Thais and it crops up in several ways in Thai folk culture. The prefix *ma* (*ma-kham*) is supposed to bring luck. Also, the branches are tough and springy: such a tree is called 'sticky' in Thai, and money is said to stick to a man with 'sticky' trees on his land. Tamarind seeds are a coppery brown; as magical properties are attributed to copper, carrying tamarind seeds is said to make the possessor safe from wounding by knives or bullets. Finally, the thorns of the tamarind tree have traditionally been used in the post-natal *yoo fai* ritual in which the mother stayed close to a fire in a very hot room for several days after the birth–the thorns were stuffed into cracks in the walls, in order to protect mother and child from evil spirits. A similar tradition, called 'mother roasting', was once also prevalent in Vietnam.

LONGAN (*Euphoria longan*)

The longan is a close relative of the litchi, grown in the cooler, highland parts of South-East Asia. Like the litchi, the longan was brought south from China by Chinese emigrants hundreds of years ago. Thailand produces and exports large quantities of canned longans–pink longan (*lamyai*) is considered to be particularly good. Longan peel is thin and brittle; picking it off is a slow task, but the succulent white flesh is ample reward. Cooked longans are also delicious and longans are dried by the Javanese and Chinese, then used to make tea.

This is another small tree, with narrow, dark green leaves. The fruits, produced in long strings on the tree, are used in folk medicine for sore eyes.

13. Langsat, fruits and leaves; engraving (Marsden, 1783).

LANGSAT (*Lansium domesticum*)

Langsat fruits hang in bunches of 8–20; the smooth outer skin is a dirty yellow colour (see Plate 13). Under the thin peel, which exudes a milky sap, are about five white or pinkish segments, unequal in size. Most segments are sweet, but one or two contain a viable seed and are very bitter; some people enjoy this contrast of flavours. A variety of langsat is the duku, with thicker peel but no milky fluid. Langsat grow on trees of medium size (up to 17 m) which take as many as 15 years before starting to bear fruit; the trees bear heavy crops twice yearly. Langsat predominates in the north of Malaysia, duku in the south and in Indonesia.

The peelings of these fruits are burned for their aromatic fragrance which is said to repel mosquitoes. Powdered langsat bark is used as a remedy for scorpion stings.

9

The Spices of the Indies

THE spice trade is an ancient one–there is evidence that cloves regularly reached China and Europe from South-East Asia 2,000 years ago. For over 200 years, from about AD 1600, European governments attempted to control the most important sections of the spice trade with measures that eventually led to the decline of the stock of spice trees and the impoverishment of the islanders who first collected and later cultivated the spices.

Spices played a role in the exploration and colonization of South-East Asia by European powers. The history of the spice trade is a story of wily traders who kept their sources secret (often for centuries) and of intervention by governments, so valuable was the trade. Lamentably, in modern times, synthetic materials have in some cases replaced the natural spices.

Some of the major spices grown in South-East Asia are nutmeg and mace, cloves, cardamoms, cinnamon, and vanilla. The spices may be dried for preservation, making them very light in weight and easy to transport over great distances without deterioration. Moreover, very little goes a long way in flavouring food. Not all the spices are fruits: nutmeg and mace come from a tree fruit and vanilla is the fruit of an orchid. Cardamoms are seeds, cloves are unopened flower buds, cinnamon is made from the bark of the cinnamon tree.

Both nutmeg and cloves are native to the Moluccas, a group of islands that lie across the equator at the western edge of the Pacific Ocean. Once known as the Spice Islands, the Moluccas now form part of Indonesia. From Roman times to the Middle Ages, spices reached Europe passing through the hands of many traders: the oar-propelled *kora-kora* is unsuitable for transporting produce in bulk, so Moluccan islanders sold their spices to Javanese traders who relied on the monsoon winds that blow out of Asia during the autumn and early winter to take them to Ceylon and the Malabar coast of India. Muslim traders took the spices further, to Aden, Cairo, and Alexandria. Italian merchants–often from Venice–carried them into the heart of Europe.

In the course of this journey, customs duties of 10 per cent were levied twelve times and profits were taken by several sets of traders, so that in Europe, cloves and nutmeg were exorbitantly expensive. Rumours of the location of the Spice Islands encouraged the Portuguese ruler, King Manuel, to send Vasco da Gama along an easterly route in 1497. The rumours also lured subsequent explorers through the Melaka Strait and the Sunda Strait, past Java, Borneo and Sulawesi to the 'disappointingly small' volcanic islands of the Moluccas. The Portuguese occupied the islands, making the spices that grow there the monopoly of the Portuguese royal house.

Sir Francis Drake reached the Moluccas during his 1577 circumnavigation of the globe and traded there; Dutch navigators followed. By the end of the seventeenth century, the Dutch were in control of this valuable trade but plant specimens were smuggled out by a French horticulturist. They were taken to the French territories of Réunion and Martinique, and the monopoly of the spice trade was eventually broken.

NUTMEG AND MACE (*Myristica fragrans*)

Nutmeg is the dried seed of the nutmeg tree, coloured mid-brown and streaked with a darker brown (see Colour Plate 19). Around the seed is a bright red, thin and patchy pulp–this is the mace, which is dried and used in savoury dishes and preserves, whereas ground nutmeg is usually added to sweet dishes, milk puddings, cakes, and punches. The green fibrous outer husk of nutmeg is also used in Malaysia, to flavour sweetmeats and jellies. Christopher Fryke described the nutmeg tree and its fruit in the seventeenth century: 'almost like a Pear tree, but doth not spread so much . . . the Fruit is much like a Peach in bigness and looks, of extraordinary fine taste and delicate smell when it is ripe.' The small flowers are not unlike lilies-of-the-valley.

The Dutch had tried to concentrate nutmeg on the island of Amboina in the Moluccas group, by cutting down nutmeg trees on other islands. However, pigeons carried the seeds back, thus thwarting the attempt. Eventually the tree was taken to other suitable tropical islands; Sir Stamford Raffles, founder of Singapore, had a hand in the introduction of nutmeg to Indonesia, by backing plantations in Sumatra. Indonesia and Grenada are now the main producing nations.

CLOVES (*Eugenia caryophyllus*)

The dark red fruit of the clove tree, called mother of cloves, is used locally as a spice, but the unopened flower buds are more pungent and give the clove-spice of commerce. A close relative of the waterapple, the clove tree was thus described by Christopher Fryke: 'The Clove Tree is much like the Laurel-Tree, the Blossom is White at first, then it turns Green and after that Red. While it is green it smells so fine and sweet that nothing can be compared to it. . . . Where the Trees grow, there grows no kind of grass or any green thing nigh . . . they draw all the moisture about them to themselves.' See the lithograph accompanying the account of Nieuhoff's travels, Plate 14.

It was traditional in the Spice Islands for parents to plant a clove tree when a child was born. The tree in some way symbolized the child, indicating his age and mirroring his health. In the seventeenth century, the Dutch tried to reduce clove

14. Clove tree; lithograph (Nieuhoff, 1704).

production by ordering all clove trees on islands away from Amboina to be cut down. This must have caused great consternation to the local inhabitants, as the death or destruction of a child's clove tree foretold the death of the child.

Used for cakes, punches and for flavouring hams in the West, much of the Indonesian clove crop is used locally as a tobacco additive, in the production of the clove cigarettes of Java. Cloves are also added to the betel quid as a flavouring. Christopher Fryke warned: 'The smell of 'em is so strong that some People have been suffocated with it when they have been busie with too great quantities of 'em and in too close a place.' Clove oil, distilled from the green parts of the tree, was long used to help soothe toothache and to sweeten bad breath. It still has uses in medicine and dentistry, such as infant teething gels, as well as in toiletries and as a perfume base. In ancient Persia and China, cloves were believed to have aphrodisiac properties.

CINNAMON (*Cinnamomum* spp.)

The spicy fragrance of cinnamon warms the chilly evenings of the northern countries when added to cakes and punches, bringing with it something of the heat and colour of the tropics. Cinnamon quills are made from the thin bark of a shrubby tree. During the rains, when the sap is running in the thin stems, the bark can be stripped off easily after two incisions have been made around a stem (only half the circumference of bark is removed, so the shrub can survive). After a few days' fermentation, the outer surface and other impurities are scraped away from the bark, which is then dried. As it dries, the section of bark curls into a quill.

True cinnamon is grown in Sri Lanka but a closely related species, also known as Padang cassia (*Cinnamomum burmanii*) is cultivated in Indonesia (see Plate 11). Indonesia is the main producer of South-East Asia's cinnamon crop, much of which is used in the region to flavour betel quids.

VANILLA (*Vanilla fragrans*)

The great orchid family provides us with myriad exotic blooms of all colours, shapes, and sizes; some with fragrance but a few with quite unpleasant smells. Only one orchid has an edible fruit, and that is the extremely valuable vanilla orchid, a native of Mexico. Because of its value in commerce, attempts have been made to cultivate this orchid in all tropical regions, but success has been very patchy. Like all orchids, *Vanilla fragrans* is exacting in the conditions in which it grows. However, some of Indonesia's many islands have proved suitable, and Indonesia now produces more vanilla than Mexico.

This orchid is a perennial vine, clinging to support with its roots. Its large, waxy flowers have pale yellowish-green petals; the fruit is bean-like, 10–25 cm long. Throughout the world, the main use of vanilla is for flavouring ice-cream but it is, of course, also very important in cake- and pastry-making.

CARDAMOM (*Elettaria cardamomum*)

Cardamoms are the seeds of a perennial herb of the ginger family, native to the Indo-Malaysian region. The plant grows to about 3 m and has aromatic seeds that become yellow when dried. True cardamoms belong to the herb layer of the evergreen rainforest and are now produced in greatest quantities in India and Indonesia. Several similar species are grown in Java and Thailand, and like many other spices, are used as flavouring for betel quids. Cardamom also flavours liqueurs and bitters, and is used in the manufacture of perfumes.

Cardamoms from South-East Asia were an important food flavouring in Roman cuisine in ancient times, imported via Alexandria. It seems probable, from early descriptions of the taste, that the Romans' cardamoms were not of the same species as those currently cultivated.

10
Fruits in Arts and Handicrafts

THE South-East Asian environment is filled with natural forms and colours which the artist and the craftsman can strive to emulate. In the decorative arts, Chinese influences from the north have emphasized flowers, birds, trees, and fish; the strictures of Islam to the south have meant that plant and abstract forms are favoured in designs rather than human figures or animals.

The traditional arts and crafts of the region are dominated by textiles, particularly by the beautiful *ikat* textiles for which either warp or weft is dyed before the cloth is woven, and by the batik cloth dyeing process used in Malaysia and Indonesia. Other important crafts are the manufacture and decoration of ceramics and the making of mats and containers by weaving rattans, palm leaves, and other fibres.

Ceramics

The fruits and the flowers of native trees have long since been adopted by the designers of the region. In the manufacture of ceramics, fruits contribute not only to painted designs but also to the shape of objects. Stylization of designs over generations means that it is frequently difficult to recognize the origins of a motif as it becomes blurred and simplified, then decorated with curlicues and details which do not occur in nature. One clearly identifiable design element often found on ceramics is the banana or plantain leaf: long and thin, rounded at the tip, and veined with a strong midrib with finer lateral veins.

Two fruit forms often seen in ceramic lidded boxes are the mangosteen (Plate 15) and the lontar. These fruit boxes often

15. Ceramic box shaped as a mangosteen fruit; Sawankhalok incised stoneware, height 7 cm; mid-fourteenth to early sixteenth century (?) (Stratton and Scott, 1981).

carry painted designs in blue and white that completely ignore the shape of the box–perhaps depicting flowers, leaves, and tendrils. So accustomed are the craftsmen to the shape, it no longer represents a fruit to them. Lidded bowls derived from the pineapple shape are not uncommon; vessels based on the starfruit are more rare: these are pentagonal jars with five flat sides. Ceramics of these styles were often manufactured in China and exported to South-East Asia to help offset the imbalance in the Chinese balance of payments caused by imports of spices, timbers, and other goods from the region.

16. Nutmeg tree design on a sarong from the Semarang region of Java; wax batik on cotton cloth, dating from the mid-nineteenth century (Tropenmuseum, Amsterdam).

Carving

In the Thai royal court, ivory was carved by court ladies to produce small boxes even more closely representing the lontar. Purely decorative ivory carvings of fruits were also made: Colour Plate 13 shows several ivory fruits, including cashew, papaya, and banana. Colour Plate 6 shows some antique Balinese carvings of fruits in wood, a craft that continues to this day.

Silver

The creation of decorative silver fruit boxes is another South-East Asian art: mangosteens, mangoes, and pumpkins are frequently sources of inspiration here (see Colour Plate 10 and Plate 7). Colour Plate 11 shows a selection of fruits made from silver filigree: a mangosteen, perhaps a sapodilla, three langsat fruits, and a pineapple.

Textiles

The weaving process inevitably prevents a close copy of natural forms except on the finest threads, so flowers and fruit—as well as other natural symbols—can only be reproduced approximately. Nevertheless, the mangosteen provides a very common element of Indonesian textile weaving for head-cloths and *hingii* (waist- or shoulder-cloths). This motif, *kembang manggis*, is an eight-pointed star derived from the mangosteen flower and is also known as the witch's foot motif. Another common woven motif is the 'tree of heaven'—the coconut palm—which is shown as a vertical trunk with great leaves equal to its height, bending along a diagonal to earth.

Batik dyeing of cotton and silk cloth—involving the application of wax to prevent some parts of the cloth from taking up dye–is chiefly a craft of Indonesia and Malaysia. Natural forms, including flowers, birds, and insects, as well as fruits, are portrayed in Indonesian batik textiles in *ceplokan* or repetitive designs. These designs have been stylized, transforming them into geometric forms, in order to comply with the *Hadith*, the Islamic book of laws, which forbids the portrayal of animal and human forms in a clearly representational manner–the majority of the populations of Indonesia and Malaysia are Muslim.

Non-geometric or *semen* designs in Indonesian batiks are characteristically a representation of creepers, tendrils, and leaves. Mangoes, grapes, and chilli peppers are often shown amidst the foliage. The design *delima wanta* represents an unripe pomegranate and is said to symbolize promise; the *pisang Bali* motif represents banana leaves and flowers (see Colour Plate 20). Another common motif, consisting of a white oval, is thought to represent the lontar fruit (see Colour Plate 9). Plate 15 shows a sarong decorated with a quite realistic representation of a nutmeg tree: the husks of the nutmeg fruits have just opened, revealing the seed and the surrounding mace. Half hidden in the foliage are birds–perhaps the pigeons that contributed so much to the spread of this spice beyond the Spice Islands.

Manuscripts

The tamarind tree has made its own very different contribution to the visual arts of South-East Asia. Tamarind seeds have a distinctive shape and colour: a rounded rectangle, black and shiny. This forms the basis of 'tamarind seed script', the old square Pali script known as *mangyi-zi* in Burmese and used by Burmese monks to write the most sacred of Burmese Buddhist religious books, the *kamawa-sa* (see Colour Plate 17). The script, which is also seen in Thailand, differs considerably from the usual rounded Burmese script. Few monks are now left who can write in this style, so it may die completely. In the *kamawa-sa*, letters are painted in black lacquer made from the gum of the South-East Asian lacquer tree (*Melanorrhea usitata*), boiled to a semi-viscous consistency and painted on to sheets of cloth lacquered in orange-red to a stiff but flexible state. The pages of the most highly prized lacquered books were made from the waist-cloths of former kings of Burma. Sometimes sheets of brass, ivory or wood have been used instead. The rows of letters are slightly more than 1 cm wide, and 6–8 lines of script make up a page.

Fruit-carving

Another, much more ephemeral craft also deserves mention: decorative fruit-carving, a distinctively Thai handicraft. Roses, chrysanthemums, water-lilies, and lotuses are carved from fruits and unlikely vegetables such as radishes. Almost all fruits are carved (except jackfruit and durian), transforming them into something unexpected or into a convenient size for eating.

The egg-shaped sapodilla, for example, is peeled to reveal the honey-coloured flesh, a zig-zag cut is made around the equator of the fruit into the centre and the two halves are prized apart; the stones are then removed. The guava requires more care because of its small, exceedingly hard seeds. Unripe guavas are peeled, then quartered; each quarter is further sub-divided, depending on the size of the fruit. Using a thin, curved knife, the part of the fruit containing the seeds is cut away from an outer crescent of crisp flesh. The crescents–eight or more–are arranged on a plate to make a many-petalled flower.

Three large fruits are commonly used for magnificent table centre-pieces: the pineapple, the papaya, and the water-melon (see Colour Plate 21). Water-melons are carved either as bowls or baskets in which either fruit salad or water-melon balls are presented. To make a bowl, the upper part of the melon is sliced off, and the rim decorated with a pattern of curves or points once the flesh of the melon has been removed. To make a basket, the water-melon is stood stem side up and two horizontal cuts are made half way up the fruit, then two vertical cuts to meet these, leaving a wedge of peel and flesh to form the handle. The flesh is then removed from the melon (though not from below the handle until last, to retain its strength). A pattern is incised on the green outer rind and the stem scar is covered with a leaf of carved rind. Sometimes the outer surface of water-melon is much sculpted to represent a pineapple or lotus flower. The 'petals' of the lotus are eased open with the thumbs whilst the fruit is immersed in water.

Papaya is carved in similar ways to make a basket for fruit or a vase for artificial flowers carved from radish and turnip. To prevent the sap that oozes out from the cut surfaces of green papaya from burning the skin, the fruit is washed repeatedly as it is carved, and rubber gloves are worn.

Popular ways of presenting pineapples include slicing horizontally, removing the flesh from the 'shell', cutting it into chunks, and then reassembling the pineapple, as though it had not been prepared. More fancifully, the pineapple shell may be turned into a candle-holding lamp, or several pineapple shells may be stacked on top of each other to make a fine palm tree, once some foliage is added.

For all types of carving, the fruit must be fresh and preferably soaked in water to improve stiffness and durability—carved fruits will keep for a day or two only. Carving knives must be kept extremely sharp and are held in a pencil grip close to the point. Fruits that discolour after exposure to the air are immersed briefly in a solution of water and lime juice.

11
Ceremonial Use of Fruits

SOUTH-EAST ASIA is rich in cultural traditions, customs, and faiths, all of which have their traditional rites, which frequently entail fruits and flowers being offered to monks, gods, and spirits. Although the peoples of the region are still predominantly agricultural, great cities have existed for many hundreds of years: religious centres like Borobudur and Angkor Wat, and centres of temporal power where kings also have required tribute–again, often using fruits and flowers as ceremonial offerings.

Provided with all the material necessities of life–food, building materials, and fuel are usually plentiful–the peoples of South-East Asia have long had ample means and time for the decoration of themselves, their homes, and their religious buildings, as well as time to intensify their spiritual lives. All over the region, great colourful festivals are mounted, resplendent with flowers and fruit.

One of Thailand's most attractive festivals is Loy Kratong, celebrated in November each year. The focus of the festival takes place in the evening, when *kratong*, little boats of banana stem decorated with banana leaves folded in the shape of lotus leaves, are placed on the waters of Thailand's many *klong* (canals) and rivers, and encouraged to float away downstream. The little boat, which takes away with it the sadnesses and regrets of the past year, should also carry a mouthful of betel-nut and the leaf of the betel vine. If it sails away without mishap, heartfelt wishes are said to be granted.

Banana-stem *kratong* are also to be seen at Buddhist ceremonies. In many Buddhist countries, it is customary for a large proportion of men and boys to join the monkhood temporarily, for as little as a week, or as long as a year; some monks, of course, stay for life. The ordination candidate, with his head newly shaven and dressed in saffron robes, is led into the temple carrying three lotus flowers. His family makes an offering of banana-stem *kratong* filled with rice, and topped with eggs.

The Balinese calendar is filled with a miscellany of festivals; one typically gay and colourful event takes place 25 days before the important Galungan festival and is given in honour of crop plants, and particularly coconut palms, in order to notify them of the approach of Galungan. So that the palms may 'participate' in the celebrations, they are bedecked with brightly coloured skirts and scarves and there is a one-day prohibition against climbing the palms.

Many South-East Asian customs involve making offerings to show respect to elders or superiors, to propitiate spirits, or to maintain family harmony. The same fruits are to be seen time and again as the basis of the gift: small bananas, opened coconuts, and betel-nuts. Other fruits will be added when the giver is feeling wealthy–rambutans, pineapples, litchis, and whatever else the season offers. Colour Plate 22 shows offerings of flowers, fruit, and rice cakes being carried to a Balinese temple. Burmese *nat* altars and Thai spirit-houses are presented with a daily offering of fruit and flowers. Similar daily offerings are made to spirits in Malaysia. It is the 'essence' of the fruits and flowers that is offered up, wafted towards temple altars by vigorous fanning.

Fruits and flowers are to be seen at all of the rites of passage through life experienced by the peoples of South-East Asia. Pregnancy and childbirth is a time particularly rich in traditional customs and rituals. Throughout the region, a thousand beliefs have been concentrated upon the new baby to ensure its survival, and coconuts figure in very many of these.

Preparations to assure the safe delivery of a healthy, handsome child begin well before the birth. Raffles recounts how, at seven months of pregnancy, a Javanese woman would wash her body with milk from a green coconut upon which the likenesses of a well-proportioned girl and boy had been carved. The mother's diet at this time was closely watched and some foods proscribed–interestingly, custom holds that palm 'cabbage' must not be eaten during this period. Immediately after the birth, a ceremonial dish made from unripe meat of a green coconut would be prepared and eaten.

Before a birth in Perak, Malaysia, the mother would have engaged a midwife to supervise the ceremonial of the delivery. Payment for these services inevitably included the gift of a betel or *sirih* box. The midwife would be charged with such duties as spitting on the new-born baby to drive away evil spirits. These spirits were also thwarted by a much perforated coconut hung over the door of the house; this was intended to confuse the spirits with its multiple exits and entrances.

In Malaysia, the new-born baby would be very rapidly introduced to the tastes of the world. Burkill tells how a grain of salt was placed inside one half of a ripening coconut then offered to the baby as the adults present counted to seven. On the count of seven, his mouth was opened to take it. Rings of gold and silver were rubbed inside a coconut, then laid on the lips of the child, invoking blessings.

It is a Balinese tradition that, following childbirth, some of the blood and afterbirth water are placed in a yellow coconut which is then wrapped in leaves of sugar palm and buried in front of the mother's sleeping quarters. A fire is built over the place where the coconut is buried and offerings are made at a bamboo altar over the spot. The Malays plant a coconut palm to mark a baby's birth and bury the afterbirth there.

But coconuts are not the only fruits involved in the magic surrounding childbirth. In Alor, eastern Indonesia, a new-born baby would be washed with juice squeezed from a banana stem, whilst the Malays prepared a drink based on the juice of the leaves of citron, indigo, and polygonum to restore the mother. Babies would be given their first taste of bananas surprisingly soon: bananas mashed or chewed by the mother are given as early as the seventh day in Perak.

An initial period of 40 days of special precautions recurs in several beliefs. Raffles noted that during the first 40 days of her child's life, the Javanese mother was recommended to eat no sugar other than coconut sugar. In Perak, on the fortieth day after his birth, a child would be presented to the Spirits of the River to ensure he would always have fish to eat. A crowd of neighbours and relatives would accompany the parents, taking many offerings with them: coconuts, betel quids, a banana flower, and a bucket made of palm leaves, as well as such foods as rice, chicken, and eggs. Four days later, the child's mother would be ritually bathed in water perfumed with herbs and lime essence, in order to dispel bad luck and to release her from taboo.

Let us follow the rituals which might have been experienced by a young man born perhaps 30 years ago in a Malaysian kampong. After the excitement of birth, the next major ceremony in the boy's life was the ceremony of circumcision. Again, offerings of coconuts, betel, and other fruits were made. In accordance with tradition, the boy was seated on a banana trunk for the operation itself, and the surgeon was given a *sirih* box with his fee.

It would be in the rituals of engagement and marriage that the betel-nut and betel quid assumed their greatest significance. From the first discussions between families to secure a marriage, to the wedding itself, betel would accompany every act: exchanged by negotiators, offered by the prospective groom to his bride and by wedding guests to the couple's families. (If the *sirih* box were to be overturned and not set up again, the negotiators would know better than to press the suit—this would signal discreetly that the girl's family did not wish for the marriage.)

In Perak, for example, an initial engagement offering usually consisted of 20–40 betel quids, two sliced betel-nuts, plus caskets containing the other spices of the betel quid: lime, tobacco, gambier, and cloves. The caskets were intertwined with garlands of flowers. More flowers bedecked offerings of fruit: apples, bananas and grapes, for example, wrapped in cloths or coloured paper. Also offered were young coconuts, their shells decorated with an incised pattern and flowers in pots made from the plaited fronds of a coconut palm.

At last, the wedding day dawns, and the young couple are enthroned as 'king and queen for the day' on a banana trunk for the ritual of *bersanding* (sitting in state). Next, the newlyweds are bathed in water perfumed with limes or coconut milk–this water is then sprayed over the wedding guests, partly in blessing, partly in fun. A paste of coconut milk, lime juice, and rice powder is used as a cosmetic to decorate the newlyweds.

Amongst the Dayaks of Sarawak, Malaysia, similar customs exist: at a marriage ceremony, a betel-nut is split into seven pieces which are placed on a tray together with seven betel leaves, seven pieces of gambier, and seven small lumps of lime. Friends and relations then enjoy the betel quids together whilst discussing the marriage laws which will bind the two partners.

The betel-nut also has its place in Dayak birth rituals: the new-born infant is sprinkled with a mixture of betel-nut and zeodary, then swaddled and laid down upon the spathe of an areca palm–truly a natural cradle for a new-born baby.

12
Plant Propagation

In natural conditions, the fruit trees and other plants described here rely on seed production to propagate themselves. The seeds are dispersed by animals attracted by the fruit. However, plants that grow from seed will not be exact replicas of either parent–which can be very inconvenient where a particularly desirable variety of fruit exists. To prevent the mixing of genes that occurs in seed production, vegetative propagation is frequently used by horticulturists. In some cases this is easy, as some plants naturally develop parts which readily grow roots and can be separated from the main plant. For example, banana plants put out 'peepers', offshoots of the main rhizome, which may be separated from it and planted elsewhere. Some trees throw up suckers from their roots and pineapples grow 'slips' just below the fruit, which may be removed and planted out. Suckers, peepers, and slips give rise to exact replicas of their parent plants, which in turn produce identical fruits.

Three common techniques for vegetative propagation are inarching, air-layering, and budgrafting. Inarching involves taking a thin flexible branch, bending it down to secure it in place under the soil, once a section of bark has been removed. Roots grow from the cut bark and the branch is subsequently severed from the main tree. Where branches are too high or too inflexible, air-layering (also called marcottage) is used: a small section of bark is removed from a young branch and root growth is stimulated by wrapping the wound in a ball of earth (see Figure 3), perhaps containing cow-dung as fertilizer. Plastic film keeps the earth ball moist nowadays, but banana leaves were the traditional wrapping. Roots will have started to grow into the ball of earth after 6–8 weeks, making the twig

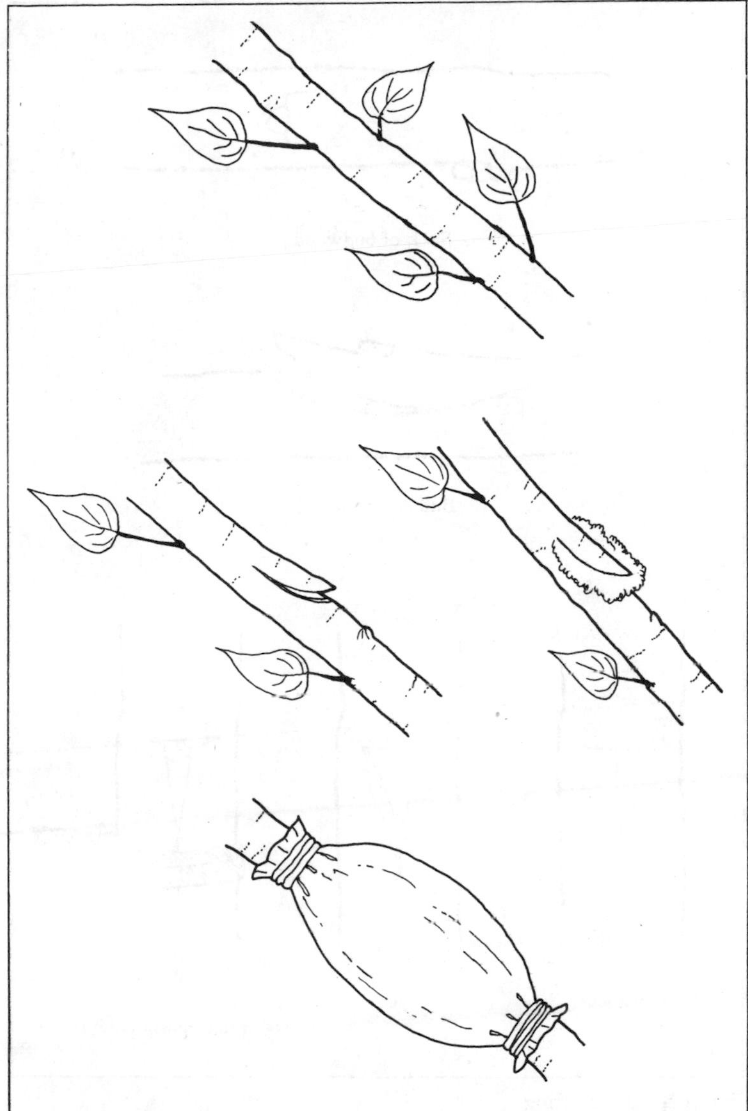

Figure 3. Air-layering

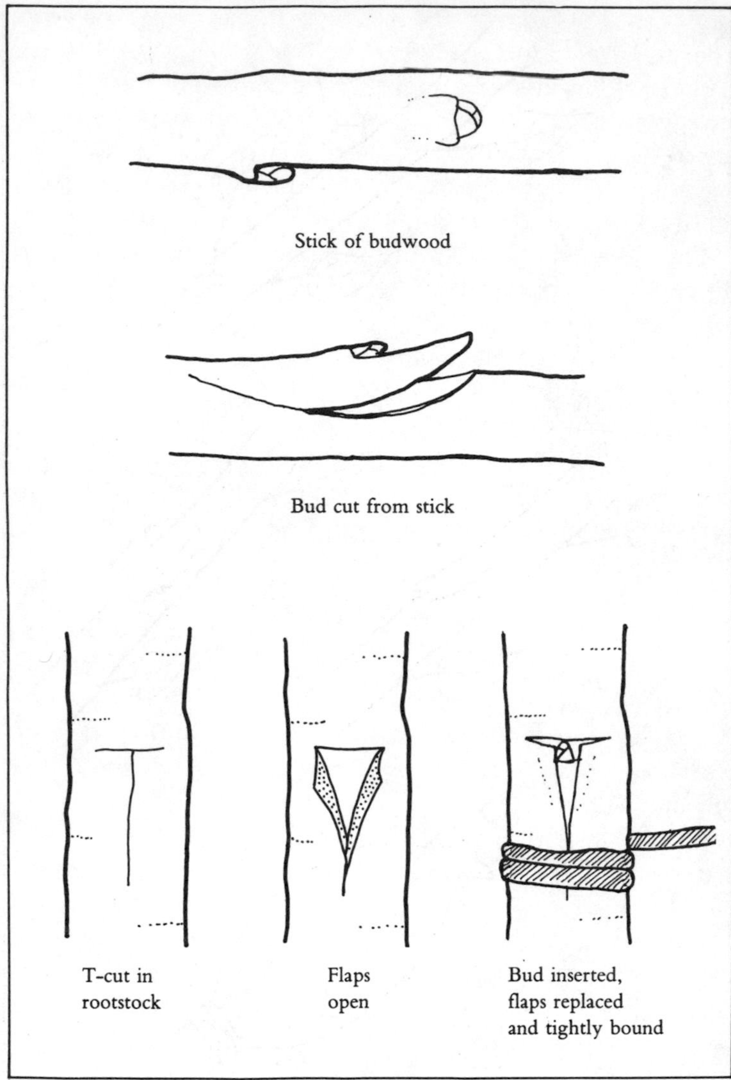

Figure 4. Budgrafting

(now called a marcot), independent. It can then be severed from the tree and planted out.

Budgrafting involves implanting a single bud from a preferred variety on to a rootstock, usually of a disease-resistant variety (Figure 4). Thus the desired characteristics of two varieties of the same species may be combined to best effect: hardy rootstock and fruit with good flavour and appearance.

These three methods enable us to copy or 'clone' the characteristics of the parent trees and their fruit, though few offspring can be produced. The revolutionary technique known as tissue culture enables thousands of new plants to be developed from a single bud, and entails carefully separating a bud into its individual cells without damaging them. The cells are then placed in a very closely controlled growing medium together with growth hormones and myriad tiny plantlets start to grow.

Home Propagation

It is possible to grow many of the species described in this book in temperate climates. Even a light frost would kill these tropical plants, so they must be grown as pot plants. Anyone with an interest in house-plants or gardening may like to try to germinate and grow some, for the sake of their foliage and the achievement itself. Two that grow quite readily are papaya and mango; the principles described below also apply to other plants.

The papaya fruit has hundreds of small black seeds, each surrounded by a gelatinous envelope inside which is a corky layer. To speed germination, wash the seeds to remove the gelatin and then prize away a little of the corky layer to reveal a small white seed inside. Sow two or three seeds in a small pot of sandy compost, and cover with 0.5–1 cm of compost. Keep the soil medium moist but do not over-water. Keep in a warm place. Within 3 weeks, shoots should appear. The first few leaves of papaya are small, tender, and heart-shaped. It is only gradually as the tree grows taller that leaf-size increases and the deeply indented shape takes form.

The mango has a very different seed which is large and enclosed in a tough outer case. As with papaya, the removal of some of the seed-case allows the seed to take up water and so germinate more rapidly–great care must be taken as the seed is fragile. The mango seed requires a larger pot than the tiny papaya seeds, and only one seed should be planted per pot, lying flat, then covered with 2–3 cm of compost. Mango germinates slowly in cool conditions but eventually the seed leaves appear above the soil. Where one seed-case is sown, two or three seedlings may come up; pinch out the weaker plants after a week or two. After the first two 'seed leaves', the leaves grow true to type: long and thin, dark green and glossy.

Glossary

Carpel. One of the flower's female reproductive organs, containing one or more ovules and giving rise to one division of the fruit, with one or more seeds.

Corm. A bulbous, swollen underground stem base which can be a means of vegetative propagation, e.g. bulb, tuber.

Floret. An individual small flower, forming part of a composite flower.

Genus. Scientific term for a group of plants or animals closely resembling each other, at a level of classification between 'species' and 'family'.

Herb. Any vascular plant which is not woody.

Inflorescence. The whole of the flowering shoot of a plant.

Mycorrhiza. A symbiotic association between a fungus and the roots of a plant, forming a layer of fungal tissue either outside the root or within the outer tissues of the root.

Nut. A one-seeded fruit which does not open spontaneously when ripe and which is protected by a shell.

Rachis (rhachis). The main stem of the inflorescence (flower stalk).

Rhizome. An underground stem which grows horizontally, living from season to season and bearing shoots and roots.

Rosette. A cluster of leaves radiating from the end of a stem.

Spathe. A large bract which ensheathes an inflorescence (flower stalk); only used for monocotyledonous plants, e.g. palms.

Species. Scientific term for a group of plants or animals sufficiently close to be able to reproduce sexually.

Totemism. The use of totems, with clan division and the social, marriage, and religious customs connected with it.

Vegetative propagation. Multiplication of a plant using vegetative (i.e. non-reproductive) parts of the plant.

Select Bibliography

Barrow, John, *A Voyage to Cochin China in the Years 1792 and 1793*, London, 1806, reprinted Kuala Lumpur, Oxford University Press, 1975.

Bastin, John and Brommer, Bea, *Nineteenth Century Prints and Illustrated Books of Indonesia*, Utrecht, Spectrum Publishers, 1979.

Beekman, E. M., *The Poison Tree: Selected Writings of Rumphius on the Natural History of the Indies*, Amherst, University of Massachusetts Press, 1981.

'Bengal Civilian', *Rambles in Java and the Straits in 1852*, London, 1853, reprinted Singapore, Oxford University Press, 1987.

Bowring, Sir John, *The Kingdom and People of Siam*, Vols. I and II, London, 1857, reprinted Kuala Lumpur, Oxford University Press, 1969.

Burkill, I. H., *A Dictionary of the Economic Products of the Malay Peninsula*, Singapore, Governments of Malaysia and Singapore, 1966.

Chin, H. F. and Yong, H. S., *Malaysian Fruits in Colour*, Kuala Lumpur, Tropical Press, 1981.

Covarrubias, Miguel, *Island of Bali*, 1937, reprinted Kuala Lumpur, Oxford University Press, 1972.

de la Loubère, S., *A New Historical Relation of the Kingdom of Siam*, London, 1693, reprinted Kuala Lumpur, Oxford University Press, 1969, and Singapore, Oxford University Press, 1986.

Fraser-Lu, Sylvia, *Burmese Lacquerware*, Bangkok, Tamarind Press, 1985.

——, *Indonesian Batik: Processes, Patterns and Places*, Singapore, Oxford University Press, 1986.

Fryke, C. and Schweitzer, C. *Voyages to the East Indies*, 1700, reprinted London, Cassell & Company, 1929.

Gittinger, M., *Splendid Symbols, Textiles and Tradition in Indonesia*, Singapore, Oxford University Press, 1985.

Hla Pe, *Burmese Proverbs*, London, John Murray, 1967.

Low, J., *The British Settlement at Penang*, Singapore, 1836, reprinted Kuala Lumpur, Oxford University Press, 1972.

Marsden, William, *The History of Sumatra*, London, 1783, reprinted Kuala Lumpur, Oxford University Press, 1966, and Singapore, Oxford University Press, 1986.

Mokhtar bin H. Md. Dom, *Malay Wedding Customs*, Kuala Lumpur, Federal Publications, 1979.

Nagy, S. and Shaw, P. E., *Tropical and Subtropical Fruits: Composition and Uses*, Washington, AVI Publishing, 1980.

Nieuhoff, John, 'Mr John Nieuhoff's Remarkable Voyages and Travels into ye best Provinces in ye West and East Indies', published in the second volume of Awnsham Churchill's *A Collection of Voyages and Travels* . . . , London, 1704. Originally published in Dutch, Amsterdam, 1682. The second part of the 1732 edition of this work was reprinted as *Voyages and Travels to the East Indies 1653–1670*, Singapore, Oxford University Press, 1988.

Purseglove, J. W., *Tropical Crops–Dicotyledons*, London, Longman, 1968.

———, *Tropical Crops–Monocotyledons*, London, Longman, 1972.

Quisumbing, E., *Medicinal Plants of the Philippines*, Quezon City, Katha, 1978.

Rajadhon, Phya Anuman, *Essays on Thai Folklore*, Bangkok, Social Science Society of Thailand, 1968.

Raffles, Sir Thomas, *The History of Java*, London, 1817, reprinted Kuala Lumpur, Oxford University Press, 1965 and 1978, and Singapore Oxford University Press, 1988.

Ramseyer, Urs, *The Art and Culture of Bali*, English edition Oxford University Press, 1977, reprinted Singapore, Oxford University Press, 1986. Originally published in German, Fribourg, Office du Livre, S. A., 1977.

Skeat, W. W., *Malay Magic: An Introduction to the Folklore and Popular Religion of the Malay Peninsula*, London, 1900, reprinted Singapore, Oxford University Press, 1984.

Solyom, G. and Solyom, B., *Textiles of the Indonesian Archipelago*, Honolulu, University Press of Hawaii, Asian Studies at Hawaii, No. 10, 1973.

Stratton, Carol and Scott, Miriam McNair, *The Art of Sukhothai*,

Thailand's Golden Age, Kuala Lumpur, Oxford University Press, 1981, reprinted Singapore, Oxford University Press, 1987.

Veevers-Carter, W., *A Garden of Eden: Plant Life in South-East Asia*, Singapore, Oxford University Press, 1986.

_____, *Riches of the Rain Forest*, Singapore, Oxford University Press, 1984.

Wallace, A. R., *The Malay Archipelago*, London, 1869, reprinted Singapore, Oxford University Press, 1986.

Index

References in brackets refer to Plate numbers; those in brackets and italics to Colour Plate numbers.